THE Mojave Road

MOJAVE ROAD

Drum Barracks to Fort Whipple

Location	Miles	Total Miles	Description
Drum Barracks	.00	.00	
El Monte	26.00	26.00	Settlement.
Crossing Rio San Gabriel . .	6.38	32.38	
Mud Spring	8.59	40.97	Stage Station.
San Jose Ranch	3.64	44.61	
Crossing Rio San Antonio	2.22	46.83	
Cocomungo Ranch	[illegible]31	[illegible]3.14	
Martin's Ranch	14.75	67.89	Ranch.
Upper Toll Gate (House) . . .	8.57	76.46	Alternately rough and sandy road.
Summit Cajon Pass	2.59	79.05	
Burton's (House)	16.08	95.13	Smooth sandy incline.
Lane's (House)	7.23	102.36	
Cottonwoods (House)	12.80	115.16	Ro[illegible]
House at Fish Ponds	17.37	132.53	
Forks of Road	12.45	144.98	[illegible]
Camp Cady	9.62	154.60	S[illegible]
Eight Mile Water	7.00	161.60	Po[illegible]ter; grass.
Caves	7.50	169.10	Water; sandy mesa nearly all the way.
Soda Lake	20.39	189.49	Water; heavy deep sand.
Point of Mountain	17.43	206.92	Heavy sand. During part of the winter the road leads round Soda Lake, making the distance four miles longer.
Marl Springs	15.98	222.90	Water; heavy sand, then rough.
Government Wells	18.48	241.38	Water; very good road.
Rock Spring	2.49	243.87	Water.
Pah-Ute Spring	18.00	261.87	Water; good road, heavy in rain; Pah-Ute Hill very bad.
Forks of Road	14.94	276.81	Left fork leads to Hardyville.
Colorado Crossing	6.62	283.43	
Camp Mojave	1.00	284.43	Headquarters, Dist. of Upper Colorado.

MOJAVE ROAD
Drum Barracks to Fort Whipple

Location	Miles	Total Miles	Description
Hardyville	6.50	290.93	Village; sandy road.
Alexander's Camp	2.41	293.34	Water and wood; no grass.
First Water Union Pass . . .	11.50	304.84	Water; grass scarce; no wood; no Camping ground.
Union Pass (Spring)	1.50	306.34	Water; no wood or grass; road up hill; no Camping ground.
Coyote Spring	16.10	322.44	Water; grass scarce; wood.
Beale's Spring	1.95	324.39	Another spring ½ mile beyond; very good water.
Hualpais Spring	14.52	338.91	Half a mile to right of road; water bad; good grass.
Tanks	12.90	351.81	Filled with sand; no water or wood.
Cottonwoods	7.90	359.71	Water, grass, and wood.
Camp Willow Grove	4.50	364.21	
Fort Rock	9.10	373.31	Water, grass, and wood. Road generally good.
Camp near Muddy Canon . .	10.91	384.22	Water in canon 300 yards to right of road; wood abundant; road good.
Anvil Rock	4.09	388.31	Water and grass.
Oaks and Willows	8.84	397.15	Water, grass, and wood. Road generally good.
Old Toll Gate	9.07	406.22	Abandoned. Road hilly, otherwise good.
Boblett's (Ranch)	2.08	408.30	Water, grass, and wood.
Toll Gate (Ranch)	1.50	409.80	Water and wood abundant. Road as above.
Williamson's Valley	14.83	424.63	Water and grass; no wood. Road excellent.
Lee's Ranch	13.21	437.84	Water, grass, and wood. Road excellent.
Fort Whipple (near Prescott) .	11.25	449.09	Road excellent.

Itinerary of the Mojave Road from Drum Barracks to Fort Whipple Measured by HENRY M. ROBERT, Major of Engineers, U. S. A., in 1867

THE

Mojave Road

by

Dennis G. Casebier

Tales of the Mojave Road
Number Five -- Sept. 1975

Tales of the Mojave Road
Publishing Company
P.O. Box 307
Norco, California 91760

Library of Congress Card Number 74-33063
ISBN 0-914224-04-2

Printed by the King's Press of Norco, California

for

E. I. Edwards

gentleman, scholar, friend

FOREWORD

Today the metropolitan area of Los Angeles, California, is connected with Albuquerque, New Mexico, by Interstate Highway 40 and the Santa Fe Railroad. There is a constant and heavy flow of commercial and private traffic between the two points over these great thoroughfares. From Albuquerque, traffic flows to and from various points in the east. From Los Angeles there are connections with the ports of the world.

The route used, starting in Los Angeles, goes eastward through the Los Angeles Basin to San Bernardino and then across the mountains by way of Cajon Pass. Entering the Mojave Desert beyond the pass, the route travels northeast for about forty miles. At this point, near the 35th parallel of north latitude, an easterly course is pursued close to that parallel all the way to Albuquerque. The specific latitude is mentioned because, in early historic times, this route was frequently referred to as the 35th Parallel Route. The distance traveled between Los Angeles and Albuquerque is nearly eight hundred miles.

This book deals with the history of a vital section of this route of travel. Special focus is given to that portion of the road across the California Desert to the Colorado River that evolved from the ancient Mohave Indian Trail. The history of that trail from the earliest times through wagon road days is treated in this book. In accomplishing this, it is frequently necessary to discuss the 35th Parallel Route as a whole and expand coverage to points far removed from the Mojave Desert.

One of the primary reasons for preparing an overview history of the Mojave Road at this time is that, hopefully, it will contribute to the preservation of the 130-mile stretch of the old road that remains substantially in its original form. This is the segment from the site of old Camp Cady (about twenty miles east of Barstow) eastward 130 miles to the point where the old road crossed the Colorado River at Fort Mojave and Hardyville. This 130 mile stretch (not including the Camp Cady Ranch and a few other points that are private land) is mostly in the public domain. With proper recognition of the history of this old trace it might well be marked out, preserved, and utilized as a hiking and recreation trail.

The reason this 130 mile segment retains much of its original form is attributable to the fact that the old Mojave Road was bypassed by twenty miles or so when the railroad was built across the eastern

Mojave Desert in the early 1880s. Later auto roads evolved from the track of the railroad leaving the Mojave Road in wagon road condition.

In many areas the old road can be followed easily today. For some miles east of Cady it followed the bed of the Mojave River so its exact course is known even though parts of the original trace have been obliterated by flooding. Along a few short stretches the original trace has been erased by drifting sand—but even in these areas it can be reestablished by surveys and historical archeology.

My hope is that the old road will be marked out clearly on the ground and managed as a recreation trail. Parts of the ancient trail would be used by foot traffic but portions might be passed over with motor vehicles. Campgrounds could be established close to the campsites used by early travelers. The springs are still there and their water could be used by recreationists now in much the same way they were used by the sojourners of that early historic period.

Whether any or all of this objective will come to pass depends almost entirely upon our governmental agencies—federal, state, and local—finding the inspiration, strength, and support to carry the project forward. In recent years studies have been conducted (particularly by the Bureau of Land Management) which have resulted in a wider recognition than existed before of the immense value of this ancient historic and prehistoric trail, and the magnificent eastern Mojave Desert through which it runs. It is now time for action or we will soon be left with neither the old trail nor the desert.

This book covers the history of the old Mojave Road from the earliest times up to 1883 when the railroad was completed along the 35th parallel. This convenient breaking point was selected because routes of travel were changed dramatically with the coming of the railroad.

I have emphasized the period before 1860. Primary focus is on the events and steps which led to the establishment of the Mojave route as a viable wagon road in November of 1859. Once that historical event has been achieved and described, the subsequent history of the wagon road is treated in more general terms. This approach keeps the book to manageable size and at the same time takes advantage of the fact that most of the material in the earlier books in this series has dealt with incidents and subjects in the period subsequent to 1859.

I have been studying the history of the old Mojave Road for over ten years. I have already mentioned my obligation to many institutions and individuals for research assistance in the previous four books in this series.

Considering its length, the present book was prepared in a short time. The schedule was compressed in an effort to assure that the book would be available during the year of the Bicentennial of the American Revolution. I could not have maintained this schedule without the capable assistance of interested friends.

Marcella Higgins Springer—a technical editor by profession—gave of her own time to read the manuscript and made a valuable contribution in minimizing the number of errors in her department.

To Sheriff Edward "Eddie" Alsip, and all my fellow-members of the San Bernardino Corral of the *Westerners*, I am indebted for expressions of confidence in and encouragement for my research and publication projects.

To Clyde F. Trudell—careful and dedicated student of the life and times of Edward F. Beale—I am indebted for a reading of the Beale parts of this work. Clyde provided valuable comments.

As noted previously, much of the country through which the Mojave Road passes in the eastern Mojave Desert is in the public domain. This resource is administered by the Bureau of Land Management. I am indebted to the following employees of that organization for official courtesies, assistance in the field, and encouragement: Wes Chambers, Bill Flint, Joe Gulliksen, Neil Pfulb, and Delmar D. Vail.

Stanford and Leila Shaw of Newberry Springs have been companions on many desert trips. Their home is on part of the old Camp Cady military reservation and through their friendship and hospitality I have learned a great deal about that locale.

From Ken and Mabel Wilhelm—proprietors of the Camp Cady Ranch—I have learned much about the Mojave Desert and about Camp Cady itself. I am greatly indebted to these interesting people whose personal adventures on the desert could easily fill a book larger than this one.

I want to extend appreciation to James L. Mayfield and the other members of the San Bernardino County Board of Supervisors for the interest they have shown in my *Tales of the Mojave Road* and for the *Resolution* passed by that body on January 13, 1975, regarding these publications.

As with my other publications, it has been my good fortune with this one to have the friendship and assistance of E. I. Edwards. This book was prepared right on the heels of *Fort Pah-Ute California* and I might well have treated myself to an extended break had it not been for Eddie's encouraging belief that this book, too, needed to be done and that it should be done now. He helped with the planning and structure of the work and then, when a draft of the manuscript was available, he read it closely several times and provided many valuable inputs.

Friends Bill and Diane Alberts devoted many evenings in coming to our home to work on this book. Once the manuscript was sufficiently developed into a relatively smooth form, they read it and provided valuable comments that have significantly affected the final result. Their *many* readings of the manuscript against typeset material have been invaluable and enabled me to publish this book in the scheduled time. The Alberts' have been companions on many desert trips—going all the way back to 1954.

To my immediate family I owe the most. The ten years spent tracking down every trace of the old road between Camp Cady and Fort Mojave have been a family experience—as have been the countless hours spent pouring over old documents, typewriters, microfilm readers, and light tables. On the present book Marlou worked with me, providing assistance and encouragement, from the time that the first word was written until the last word was proof read and the book finally "put to bed."

Dennis G. Casebier
September 23, 1975

CONTENTS

ILLUSTRATIONS

MAPS

Winfield Scott Hancock
Courtesy, National Archives

INDIAN TRAIL

Long before the coming of white men an ancient trade trail stretched across the Mojave Desert. It connected the villages of the Mohave Indians on the Colorado River with the Pacific Coast—a distance of nearly three hundred miles.[1]

The Mohaves were agriculturists. Each year—usually in June—the Colorado was raised by melting snow from the distant Rocky Mountains. The river rose over its banks and flooded narrow bottoms. After a few weeks—sometimes within days—the water receded. The Mohaves planted beans, squash, melons, corn, and wheat in the bottoms that had been flooded. This, with nature's contribution of mesquite beans, supported the powerful Mohave Nation of perhaps several thousand souls. If the river failed to overflow the Indians faced famine.

Sea shells were held in high regard by the Mohaves as they were with most American Indians. The sea was nearly three hundred miles distant—but the Mohaves had the desire, the strength and, being agriculturists, the time to make the trip. They were well-known as travelers throughout much of the area that is now Arizona, Nevada, and southern California. To trade for the shells they brought pottery, gourds, dried pumpkin, mesquite beans, and other food products. They also acted as middle-men bringing blankets and other goods manufactured by the Hopis to trade with the coastal and mountain Indians far to the west of the Mohave Villages.

It may be wondered how these trading expeditions could have been justified from a purely commercial point of view. Without beasts of burden, how much could these traders have carried either to or from their distant market places? A white girl who was for a time a captive among Apaches and later among the Mohaves, provides us with a little insight on this point. When among the Apaches she noted that Mohave traders visited about once each year the village where she was held captive. The distance from the Mohave Villages was probably over one hundred miles. She described the first visit to the village as follows: "During the autumn of 1851, late in the season, quite a large company of Mohaves came among us on a trading expedition. But the whole transactions of one of these expeditions, did not comprise the amount of wealth or business of one hour's ordinary shopping of a country girl."[2] From this statement and from other hints in early accounts, the impression is gained that the business-end of the long jaunts the Mohaves took was small. It seems clear that the gain to the individual Mohave was probably not as much in the material he brought back home as it was in the adventure in which he had been involved.

"Mojave Indians" as they appeared to the Whipple railroad survey party in 1854. From: *Reports of Explorations and Surveys, to Ascertain the Most Practicable and Economical Route for a Railroad from the Mississippi River to the Pacific Ocean* (Vol. III). By Lt. Amiel W. Whipple and others.

Mohave runners made the trip to the coast in just a few days. The first one hundred miles west of the Mohave Villages was the heart of the eastern Mojave Desert. Travelers were dependent upon the natural springs and seeps that occur in this desert land. They could carry only small quantities of water in gourds. Exact routes through this country varied, depending upon the availability of water which changed with the seasons and rainfall. But all routes led to the sink of the Mojave River—the dry lake that became known to white men as Soda Lake.

The Mojave River is a strange stream. Its importance to the Mojave Road, and hence to this book, is so great that it justifies special mention here. Its unusual characteristics have earned it the name "Upsidedown River." The following properties prompted this name. First, it is larger at its source than at its ending. In fact, the farther it is away from its source, the smaller it is. Second, it flows away from the ocean instead of toward it. Its waters never reach the sea. Third, throughout most of its course it flows underground. Once away from the mountains its waters rise to the surface only occasionally along its course.[3]

It is the mountains that create the desert; it is the mountains that also create the strange Mojave River. The moisture-laden clouds from the ocean deposit their rain and snow in the chains of high mountains that run north and south through California. Inland from those mountains is, as Mary Hunter Austin so feelingly puts it, *The Land of Little Rain*—the great deserts of California.

It is understandable that some of the rain and snow from those moisture-laden clouds would fall on the desert side of these high mountains. The water so deposited runs out into the desert giving rise to several interesting and unusual rivers of which the Mojave is one. Owens and Whitewater rivers are two other examples.

The Mojave River rises on the northern and eastern slopes of the San Bernardino Mountains. It runs north into the desert for about forty miles and then turns east and runs for another seventy miles when it finally reaches its sink at usually dry Soda Lake. The headwaters of this river are close to Cajon Pass which is one of the best passes from the coastal to the desert area of California. Long ago the Mohaves learned that the Mojave River provided a stream of life-giving water for more than one hundred miles out into the desert and that the mountains were easy to cross near the headwaters of the Mojave. These are the main geographical features that determined the route of the ancient Mohave Indian Trail.

"Dwellings of the Natives of the Rio Colorado of the West." A scene in Mohave Valley drawn by artist Heinrich Baldwin Mollhausen who visited the valley with Whipple in 1854. From: *Diary of a Journey from the Mississippi to the Coasts of the Pacific with a United States Government Expedition* (Vol. II), by Baldwin Mollhausen.

SPANISH EXPLORERS

Our understanding of the Mohave Indian Trail owes much to the early explorer Fr. Francisco Garces. He was probably the first white man to visit the homeland of the powerful Mohaves. Also, he was perhaps the first to cross what we now call the eastern Mojave Desert and to use the Mojave River trail across the desert to the San Bernardino Mountains.

When Father Garces made his famous trek in 1776 the Mohave Indian Trail was in essentially prehistoric form. The Mohaves knew of white men but the presence of these strangers had not yet impacted importantly upon their way of life. The Indians living out on the desert itself—the Pah-Utes—can be considered to have been in prehistoric form also.

Francisco Tomas Hermenegildo Garces was a Franciscan assigned to the mission San Xavier del Bac near the present site of Tucson, Arizona. He had accompanied Capt. Juan Bautista de Anza from Arizona overland by a southern route to the San Gabriel Mission in California in 1774. On September 1 of the next year he departed from southern Arizona again, this time as part of Anza's famous expedition which was destined to establish Spanish settlements at the Bay of San Francisco. The expedition reached the Yuma Villages on the Colorado River in early December. The main party soon pushed on to California, but Garces and a few others remained on the Colorado among the Indians. A short time later Garces set out up the Colorado with only Indians for companions on one of the most noteworthy explorations of the American West ever undertaken. He traveled first up the Colorado River to the Mohave Villages, thence overland to San Gabriel Mission and elsewhere in southern California, thence back to the Mohave Villages, thence eastward across most of Arizona to the Hopi Villages. From Hopi he returned to the lands of the Mohaves and then traveled back down the Colorado to the Yuma Nation and on to San Xavier del Bac. Near the beginning of his travels Garces explained his mission as follows:

> I was directed, together with another religious, to join Lieutenant-Colonel Juan Bautista de Anza and the Rev. Father Pedro Font, who were to go to the harbour of San Francisco, accompanying them as far as the Colorado River, where with my companion I was to await their return and in the meantime to look over the country and treat with the near-by Indian nations, determining if they were disposed and ready for religious instruction and for becoming subjects of our Sovereign.[4]

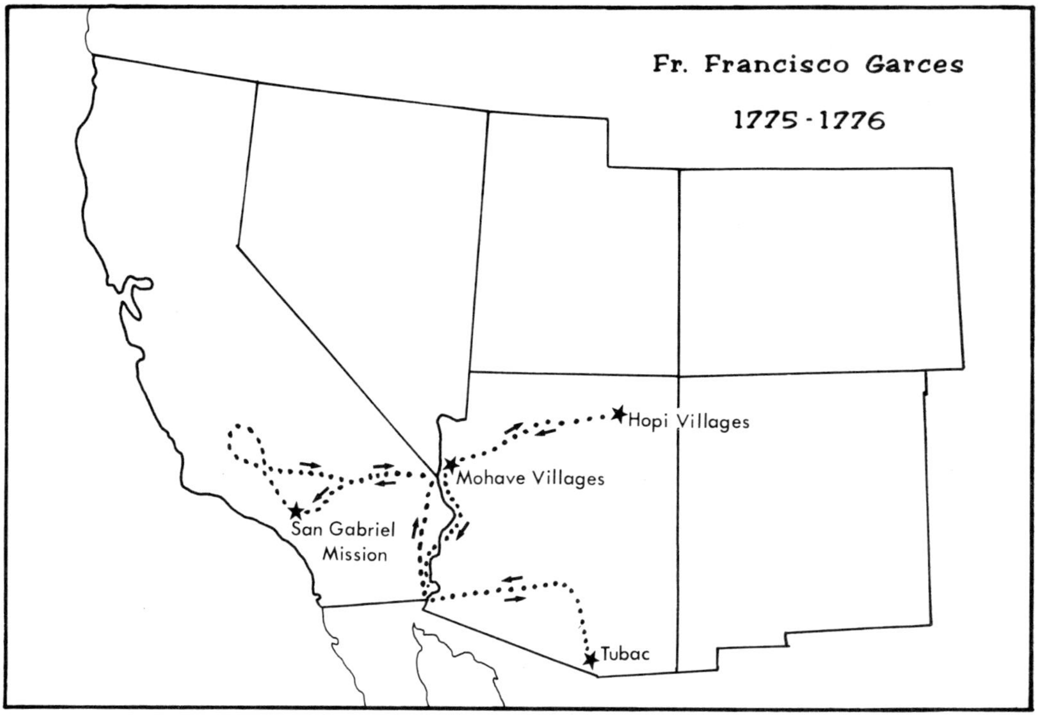

In February of 1776 Garces reached the territory of the Mohave Indians—or *Jamajabs* as he called them. At this point he decided to travel on directly west in an attempt to reach the Spanish settlements on the coast. In his diary he explained: "I laid before them [the Mohaves] my desires to visit the Fathers living near the sea; they gave assent and offered to accompany me, for they had heard of them and knew the way."

Later in his diary, Garces explained the reason for making this trip: "My principal object on leaving the Jamajabs' country was to go straight to San Luis [Obispo] Mission, or further north, to make communication easier between the Sonora and Moqui [Hopi] provinces and Monterey (which is what His Excellency the Viceroy seeks)... ... ..." Hence, Garces was exploring the route that would later became the 35th Parallel Route, or Beale's Wagon Road, or—even later—the Santa Fe Railroad and U.S. Interstate Highway 40.

Whatever his reasons, it is fortunate for those of us who are interested in the historical lore of the eastern Mojave Desert that Garces determined to make this trek. He crossed our desert twice in the first half of 1776 and maintained a diary in sufficient detail that his general course can be identified. Of greatest value to this book is the fact that Garces recognized and understood the importance of the Mohave Indian Trail. In a number of places he described the trail and the traffic over it. His early accurate assessment provides the basis for understanding later references to the route. The allusions to the Mohave Indian Trail and the trade conducted over it have been extracted from the Garces diary and they are repeated here.[5]

> Here [somewhere near the Providence Mountains] I met four Indians who had come from Santa Clara to traffic in shell beads. They were carrying no food supply, nor even bows for hunting. Noticing my astonishment at this, where there is nothing to eat, they said, 'We Jamajabs can withstand hunger and thirst for as long as four days,' giving me to understand that they were hardy men.
>
> There came to it [a point on the Mojave River] five Jamajab Indians who had just returned from trading at San Gabriel, well pleased at their treatment by the Fathers, who had given them maize. These Indians imitated the bleating of the calves.
>
> Some Indians from the north-northeast arrived [Garces was in the Santa Clara Valley at this time] who wanted to take me to their country with five more Jamajabs who had come to trade.
>
> I went to the rancheria of San Pasqual [somewhere in the present Kern County], where I found two Jamajabs who had recently arrived (of those who had journeyed with me there remained only Luis and Ventura); they were evidence to me of the regular trade between these nations and those by the sea.
>
> I delayed here [probably somewhere near the Providence Mountains in eastern San Bernardino County once again–this is on the return trip to the Mohave Villages] because some of the Jamajabs who had come to trade in shells were ill.

These brief descriptions show the nature and purpose of the old Mohave Trail as it existed in prehistoric times. For many years after Garces the trail would continue to function. Very soon, however, the Mohaves would begin having trouble with the Spaniards and later with the Mexicans. Garces himself seems to have foreseen this in 1776 when he was at the San Gabriel Mission. He wrote as follows: "A little before my arrival, there had been at San Gabriel for their traffic in shells a few Jamajabs (those I have already said I met). When news of them reached the Commander [who was in San Diego], he wrote to the corporal at this mission to seize these Indians and take them toward their land a long way from here–an order that was not carried out because it arrived after they had left. ... ... The law of nations permits trade between one people and another. What reason, then, can there be to stop the harmless and long-established commerce of the river people with those of the sea, consisting as it does in some white shell-beads?

If we preach to the heathen a law of peace and charity, how can we think of sowing discord?"[6]

Garces may have considered that the Mohaves constituted an autonomous nation—and hence its citizens entitled to the benefits of the law of nations—but, even if he did, few other white men entertained similar views. Indians were creatures with no more rights than the grizzly bears that roamed the foothills of California's mountains. It was inevitable that these free spirits from the great river beyond the deserts would be seen by the mission fathers as a source of trouble. Mission Indians were difficult to control and the problem was aggravated in many ways by the occasional arrival of Mohave warriors from the desert.

It would appear that productive communications were never established between the Mohave Indians and the white men in the mission establishments. All the white men knew was that somewhere beyond the mountains and deserts in an inaccessible place there lived a powerful nation that they stood no chance of controlling. Warriors from this nation would appear unexpectedly and spawn discontent among the Indian neophytes and steal mission cattle. Records from this period are sparse. It is difficult even to understand the Spanish side of the story—and it is perhaps impossible to understand the Mohave side of it.

Expeditions were at times sent from the mission establishment against the Mohaves. It appears certain, however, that no body of Spanish soldiers was ever able to cross the eastern Mojave Desert and strike the Mohave homeland. Perhaps Lt. Gabriel Moraga, who led a force of about fifty men down the Mojave River late in 1819, came the closest. Fr. Joaquin Pasqual Nuez accompanied Moraga's expedition as diarist and minister. Regarding the purposes of the expedition, Nuez wrote as follows: "... ... ... to punish the bold and evil deeds of the Indians of the Rancheria Amajaba [Mohave], who in the Mission of San Buenaventura to the number of twenty-one men, in the Passover of the Holy Spirit, during the Holy Sacrifice of the Mass, slew the corporal of the guard named Rufino Leiba and a retired soldier of the Presidio of Santa Barbara, Mariano Cota and a neophyte of said Mission. The Indians of the Rancheria of Amajaba had the audacity to come and commit deeds of violence on the pagan Indians of the Rancheria of Atongaibit and others of the region, and the Christians who fled from the Mission of San Fernando and this one of San Gabriel."[7]

Nuez' descriptions of places are not sufficiently detailed to determine exactly where the expedition camped or how close it came to the Mohave Villages. It seems quite clear, however, that the party proceeded down the Mojave and intended to follow the Indian Trail to the Colorado River. High up on the Mojave River Moraga acquired an Indian who could guide them over the route. Nuez reported: "A little before celebrating said function [a religious function which Nuez describes] a pagan Indian from the Rancheria of Argaipaba presented himself, who admitted that seven days before he had left Amajaba, and because he considered it ["it" refers to "traveling to Amajaba–or the Mohave Villages"] a very practicable thing to do, the commander [Moraga] determined that he should travel with us, serving as a guide."

It appears that Moraga traveled down the Mojave River to the vicinity of Soda Lake. The expedition was probably in that neighborhood on December 2, 1819, when Nuez wrote of their plans as follows: "On account of information obtained from the Indians that the waters ahead were very scarce until one days journey before reaching the Mohave settlement, the Commander of the Expedition determined to advance with ten soldiers and four of the Indians came with us." But, even with the party reduced to this small number traveling light, Moraga was stopped by the desert. Nuez went on to report the failure: "After having traveled all day and a large part of the night he was made heartsick because he was unable to proceed farther because the horses that they carried along had given out, notwithstanding that they were the strongest." And so Moraga turned back probably still at least seventy and possibly eighty miles short of his objective. The Mohaves were secure in their stronghold!

Jedediah Strong Smith

This portrait of Jedediah Smith was done by George Mathias for the Jedediah Smith Society. The portrait was painted from research done by the staff of the society on the physical appearance of this famous man of whom there are no photographs. *Courtesy, Jedediah Smith Society, Pacific Center for Western Historical Studies, University of the Pacific.*

Jedediah Strong Smith was only twenty-seven years old in 1826 when he became the first American to reach California overland. His career as a trapper had commenced in 1822 while in the employ of William H. Ashley. In the summer of 1826 Smith, in partnership with David E. Jackson and William L. Sublette, bought Ashley's interests and formed their own company.

In August of 1826 the bulk of the new company left Cache Valley in Utah and headed north for the season's hunt. But Jedediah with sixteen other men decided to head south and west from Great Salt Lake to explore that unknown area for possible new beaver country. Apparently he had no intention at that time of journeying on to California.

Passing through Utah, Smith arrived at the Colorado River in early October at the mouth of the Virgin River. He crossed and traveled south down the east side of the Colorado (although possibly a few miles distant to avoid the rough country adjacent to the river) and arrived at the Mohave Villages. Smith found the Mohaves to be friendly, apparently contented, and with ample provisions. He stayed among them for about two weeks, and in that time learned that the Mexican settlements on the California coast were not far and that the Indians knew the way there. He determined to risk the wrath of the Mexican authorities and to go there to replenish his supplies.

In a letter written the following year to Gen. William Clark, Superintendent of Indian Affairs, Smith explained the factors which prompted him to go to California and his adventures in crossing the eastern Mojave Desert. The following quotation is extracted from that letter. It provides Smith's brief description of the desert crossing.

> I travelled a west course [from the Mohave Villages] fifteen days over a country of complete barrens, generally travelling from morning until night without water. I crossed a Salt plain about 20 miles long and 8 wide [Soda Lake]; on the surface was a crust of beautiful white salt, quite thin;–Under this surface there is a layer of salt from a half to one & a half inches in depth;–between this and the upper layer there is about four inches of yellowish sand.[8]

From Soda Lake, Smith's guides led him to the Mojave River which stream Smith called the "Inconstant River" because of its peculiar habit of running underground most of the time but occasionally rising to the surface. His name for the Mojave River gained some recognition and appeared on maps for a few years. Following the "Inconstant River" to its head Smith crossed over the San Bernardino Mountains and went on to San Gabriel Mission.

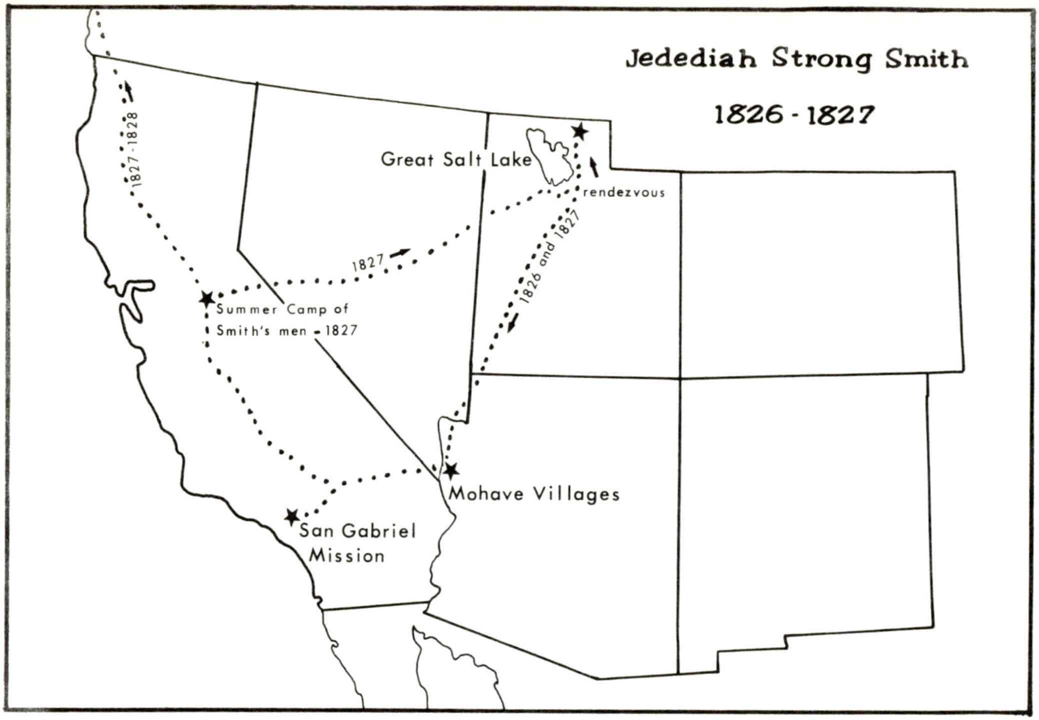

After settling his difficulties with the Mexican authorities (by this time it was January of 1827) Smith was supposed to leave California by the same route over which he had entered it. He followed his former trail over the San Bernardino Mountains but once out of observation by the authorities he turned west and ultimately headed north through California's Central Valley trapping beaver along the way.

By April, Smith's party was on the American River with a large quantity of beaver fur. They were ready to return to Utah for the summer rendezvous but his party could not cross the Sierras with their cargo because of heavy snow. Smith left most of his party with their catch on the Stanislaus River while he and two companions pushed on across the mountains to Utah. Early in July they joined the summer rendezvous at Bear Lake.

After only a short stay at the rendezvous Smith with eighteen men again left for California to join the party they had left there and to obtain their cached furs. Smith followed his former trail down through Utah and by mid-August had arrived back at the Mohave Villages on the east side of the Colorado River.

As was the case during the previous year the Mohaves appeared to be friendly. For several days he traded with them and rested his horses. On about August 18, 1827, he made arrangements to cross to the west bank of the river. At the crucial time when part of Smith's party was on one bank and part in the process of crossing to the other bank, the Mohaves attacked without warning. On the east bank there were ten men and two women. The ten men were killed and the two women were taken captive.

For the nine men remaining on the west bank (of which one was wounded) the obstacles in the way of a successful escape looked formidable. They were besieged by hundreds of Indians thirsting for their lives and once out of the clutches of the Indians they had to face first the rigors of a desert crossing and then Mexican officialdom in California. But with the determination and courage characteristic of Mountain Men, they prevailed in the face of all these obstacles. They held off the Mohaves until dark and then retreated westward into the eastern Mojave Desert. Traveling mostly at night and without the benefit of the Indian guides used during the previous year, they managed to find the springs and the "Inconstant River" and reach the Mexican settlements. Smith himself has provided a description of their trek across the eastern Mojave. In the following extract Smith describes Soda Lake and the lower part of the Mojave River:

> My men were much discouraged, but I cheered and urged them forward as much as possible and it seemed a happy providence that lead us to the little spring in the edge of the Salt Plain, for there was nothing to denote its place and the old trail was filled up with the drifting sand. Two of the men had been obliged to stop two or three miles before we got to the spring, and although it was just night two of the men took a small kettle of water and went back, found and brought them up. After dark we proceed(ed) on across the Salt Plain and stopt at the holes I had dug when I passed before and there remained for the rest of the night.
>
> On the following day I moved on to Inconstant River. It was still dryer than when I passed the year before. I think it reasonable to suppose that the Salt of the Plain has been formed by a deposit at different times from the overflowing of Inconstant River. The water of the river is sufficiently brackish and the country near the place where it is finally lost in the sand is sufficiently level to justify the conclusion that in some seasons of the year, when the water is most abundant, it spreads over the Plain, and as the dry season approaches the water disappears and leaves a deposit of salt which has in the course of years produced the beautiful encrustation found in the Salt Plain.[9]

In California, Smith met with even more official difficulty than the year before. At length, however, he was permitted to proceed on his journey. He sold the furs he had cached the previous season and with the proceeds purchased provisions for his destitute band.

Thus ends Jedediah Strong Smith's association with the Mohave Indian Trail. Smith would never cross over the route again—but the massacre of more than half of his party by the Mohaves on August 18, 1827 (a massacre that is considered to be one of the worst disasters in the history of the American fur trade), would have an important effect on the use made of the Mohave Trail by white men. Previously only missionaries on the California coast were familiar with the powerful Mohaves and their troublesome ways. But now, word of the supposed treachery of this tribe had spread throughout the American West. A direct trail would very likely have developed between the settlements on the upper Rio Grande in New Mexico and California by way of northern Arizona and the Mohave Indian Trail, had it not been for a fear of the great Mohave Nation. And much of this fear can be traced to the act of treachery perpetrated against Smith's party.

The question about why the Mohaves attacked Smith's party naturally arises. The answer may be that between Smith's first and second visits the country of the Mohaves was crossed by another party of trappers from New Mexico under Ewing Young and the Indians were badly defeated in a contest with them. Smith's arrival in August provided the Indians with an opportunity for revenge.

Leaving New Mexico in 1826, Young's party had trapped the Gila and Salt Rivers in Arizona late in 1826 and early in 1827. They then trapped down the Gila to its junction with the Colorado and up that river until they encountered the Mohaves.

According to James Ohio Pattie, who at this point was a member of Young's party, the difficulties with the "Mohawa" Indians progressed as follows: a Mohave chief demanded a horse as the price of the beaver the trappers had taken from the river. [10] The trappers refused. The Indian shot an arrow into a tree. Young shot the arrow in two with his gun and that ended the matter for the first day. The next day the Indian returned and repeated his demand. Again it was refused. The Indian started off at a gallop and on his way fired a spear into a horse belonging to the trappers. He was immediately shot dead. Thus commenced hostilities.

The next day a fight took place and Pattie reported that sixteen Indians were killed. A few nights later, while the trappers were sleeping, the Mohaves sent a shower of arrows into their camp and killed two white men. The next morning eighteen of the trappers took to the trail of the fleeing Indians. They caught up with them late in the day and, according to Pattie, they killed the greater part of the band. The final chapter in the incident was reported by Pattie as follows: "We suspended those that we had killed upon the trees, and left their bodies to dangle in terror to the rest, and as a proof, how we retaliated aggression." The chronology of these events is not definitely established, but it seems likely that they took place in March of 1827. Smith's unsuspecting party arrived in August of that year unaware that the Mohaves were seething with rage because of the treatment they had received at the hands of Young's party.

A few years later, in the winter of 1829-1830, a party crossed the eastern Mojave Desert after proceeding from New Mexico across northern Arizona. This party of trappers under Ewing Young included young Christopher "Kit" Carson. The following information regarding this event is from Carson's autobiography. The quotation commences at the point where the trappers had reached the Colorado River from the east:

> It can better be imagined, our joy, than described when we discovered the stream [the Colorado River]. We had suffered greatly for want of food. We met a party of the Mohave Indians and purchased of them a mare, heavy with foal. The mare was killed and eaten by the party with great gusto; even the foal was devoured. We encamped on the banks of the Colorado three days, recruiting our animals and trading for provisions with the Indians. We procured of them a few beans and corn. Then we took a southwestern course and in three days march, struck the bed of a stream [Mojave River] which rises in the coast range, has a northeast course, and is lost in the sands of the Great Basin. We proceeded up the stream for six days. In two days after our arrival on the stream we found water. We then left the stream and travelled in a westerly direction and in four days, arrived at the Mission of San Gabriel.[11]

The party left California to return to New Mexico in September of 1830. Carson described their trek from the vicinity of San Gabriel Mission to the Mohave Villages on the Colorado River as follows:

> The next day we departed and pursued nearly the same route by which we came, and in nine days we arrived on the Colorado. Two days after our arrival, on the Colorado, at least five hundred Indian warriors came to our camp, They pretended friendship, but such a large number coming, we mistrusted them, and closely watched their maneouvers. We discovered where they had their weapons concealed, and then it became apparent to us that their design was to murder the party. There were but few of us in camp, the greater number being out visiting the traps. I considered the safest way to act was not to let the Indians know of our mistrust and to act in a fearless manner. One of the Indians could speak Spanish. I directed him to state to the Indians that they must leave our camp inside of ten minutes. If one should be found after the expiration of that time, he would be shot. Before the expiration of the ten minutes, everyone had left.[12]

Apparently the trappers avoided further difficulty with the Mohaves. They then trapped down the Colorado River to tidewater and eventually returned to New Mexico by way of the Gila River Route.

The next party known to come among the Mohaves was Peter Skene Ogden's trapping expedition to California in 1829-1830. Ogden was leading a party of Hudson's Bay Company's trappers to California to tap beaver resources there before they could be harvested by American trappers known to be interested in the area. Ogden had discussed the route with Jedediah Smith and had the story of the massacre of part of Smith's party direct from Jedediah himself. Consequently, Ogden anticipated trouble with the Mohaves and watched closely for signs that would indicate it might be developing. Of his encounter with the Mohaves Ogden wrote as follows:

> On the day following our appearance among them [the Mohaves], they swarmed about the camp, every man carrying, in addition to his proper arms, a long stick on his shoulder, in derision of the manner in which we carry our guns. Observing the greatness of their numbers, I took the precaution of posting an extra guard over our horses, and warned the men to hold themselves in readiness for the worst. Besides their usual firearms, I furnished each of our little party with a spear, giving orders not to reload after the first volley but to charge; for I was apprehensive lest, during the interval of loading, the Indians might make a rush and overpower us; and that a speedy attack was meditated, I could no longer doubt. Our preparations completed, I admitted a few Indians into the camp, purposely that they might observe our state of defence, and with the hope it might deter them from attacking us. Unhappily for them, the desired effect was not produced, for presently one of the guard was wounded, and the alarm given that the Indians were securing our horses. This was sufficient for me. They had shed the first blood, and I was resolved that theirs should repay it; and as it was now for life or death with us, I ordered a general discharge, to be followed up by a charge with the spear. The first, however, sufficed; for on seeing the number of their fellows who in a single moment were made to lick the dust, the rest ingloriously fled, and we saw no more of them. Twenty-six remained dead on the field.[13]

OLD SPANISH TRAIL

The encounters between trappers and Mohaves in the years 1826-1830 firmly established for these Indians a reputation for hostility and treachery in the minds of Americans. This, combined with their large numbers and their strategic location along a fifty-mile stretch of the Colorado River, made them an important factor in the development of routes of travel for the next thirty years. This effect was particularly important during the early 1830s when a commercial trade route was established between New Mexico and California.

The logical path for such a route would have been to leave civilization at the Mexican settlement at Albuquerque and run to the Indian pueblo of Zuni. Beyond Zuni the route would strike the Little Colorado River and follow down that stream to a suitable point where it would deflect to the left, heading west to pass to the south of the San Francisco Mountains near where Flagstaff, Arizona, is now located. From there it would travel to the vicinity of the present Kingman, Arizona, following approximately the future track of the Santa Fe Railroad. From Kingman (or Beale's Springs, as that site was known in early historic times) the path would head directly west and strike the Colorado River in the midst of the Mohave Villages somewhat to the north of the present Needles, California. Crossing the Colorado River at that point, the Mohave Trail would be followed across the desert to the Mojave River and the mountains could be crossed by way of Cajon Pass.

Without a doubt this was one of the best natural routes for a wagon road across the American West. Certainly it would have been the logical route to connect the settlements on the upper Rio Grande in New Mexico with the Los Angeles area. The distance to Los Angeles from Albuquerque by that route was only about eight hundred miles. Instead of using this direct route, however, the first route to be developed, the "Old Spanish Trail," followed a circuitous path that led from New Mexico northwest into central Utah and then southwesterly down through Utah, southern Nevada, crossed the Mojave Desert and entered the Los Angeles Basin through Cajon Pass. The length of the trail by this route was about twelve hundred miles—at least fifty percent longer than by the direct route.

The inducements to travel that extra four hundred miles had to be strong and it would seem they could be clearly identified now. Any study of the Old Spanish Trail itself would of necessity address this important point. The definitive work on this subject is LeRoy R. Hafen and Ann W. Hafen's classic study *Old Spanish Trail Santa Fe to Los Angeles.* About the question of why the Old Spanish

Trail took such a roundabout northern route, the Hafens wrote as follows:

> Persistent hostility of the embittered Yumas, Hopis and Apaches blocked the way to achievement of the Garces' plan for a direct route across Arizona from southern California to New Mexico. Add to this obstacle the difficulties of terrain—long stretches of waterless desert, the deep forbidding gorges of the Grand Canyon country—and we have the reason why a direct line was not established and why practical travel followed the general course of the Escalante trail and traders' route into the Ute country north of the Colorado River's canyon area. This pathway, continually extended, was to be pushed forward until it finally reached the Pacific. It would become the first route of regular travel between New Mexico and California and be labeled the Old Spanish Trail.[14]

I feel certain the Hafens are correct that Indians were an important factor in the selection of such an out-of-the-way route. However, I believe the Mohaves were the main threat and not the tribes named. The Yumas were not on the direct line between the settlements on the Rio Grande and southern California and did not pose a threat to traffic on the route. The Hopis were concentrated then—as they are today—in their villages in northeastern Arizona and the direct route referred to would have missed their villages by fifty miles or so. They may have posed an occasional threat as raiders, but this would not be comparable to the dangers to be encountered if the trail involved passed directly through their country. The point about Apaches is correct only if the word is taken in the broadest sense. The tribes usually thought of when the name "Apache" is used are those living in central or southern Arizona, well south of the line of this proposed direct road. The most powerful tribe of "Apache" type Indians traversed by this direct route would have been the Navajos and next in line for this honor were the Yavapais and Walapais. These three tribes would have constantly annoyed passing trains (as they later did when wagon traffic commenced passing through this latitude) but they would have been no real threat to a well-armed and alert party.

It seems to me that the Mohaves must be considered as the most important element of the Indian threat that was avoided by taking the circuitous route up through Utah. The Mohaves were situated on both sides of the Colorado for a distance of fifty miles and there was no way their country could be avoided. They were numerous; and, not being nomadic hunters, they were always concentrated "at home" so there was no hope that the traveler could slip by unnoticed. Also, while in the midst of this numerous and warlike band, the traveler must cross the Colorado River. It is my impression that the crossing itself was not a thing that the average experienced traveler feared (at least not to the extent of being willing to travel more than four hundred miles out of his way to avoid); but, when it had to be done in the presence of a large warlike tribe, it created an unacceptably dangerous situation.

Undoubtedly, the fact that the Mexicans were on friendly terms with the Utes and had traded in their country in Utah for many years was an influence in selection of the Utah route for the Old Spanish Trail.

So far as difficulty of terrain is concerned, the route to the north crossed more difficult country than the direct line across northern Arizona would have traversed. True, the canyon country of the Colorado would have been close at hand in northern Arizona, but not many miles south of the Grand Canyon itself the country is not broken. U. S. Highway 66 between Kingman and Flagstaff passes through some very spectacular scenery, but the country does not include many deep forbidding gorges nor much badly broken country. Also, there was abundant water, wood, and grass along this direct route, although the exact locations of these essential articles would have to be determined and learned, as along any other route.

It is generally considered that the Old Spanish Trail, as a commercial trade trail, was inaugurated in 1829-1830 with Antonio Armijo's trip from New Mexico to California. The expedition left New Mexico in November of 1829. The following quotation explains the motivation behind the trip:

> On the 6th of November of the past year, there left from the village of Abiquiu 31 men, including the commandant Citizen Antonio Armijo, inhabitants of the territory of New Mexico, wishing to discover a route to Upper California and to sell therein some manufactures of their country, traveling towards the Northwest, and a month later reaching the Rio Grande, or Colorado, which they forded without difficulty, despite its being about 2,000 *varas* wide, and on the banks of the said river, which are of smooth stone, there are some inscriptions which they inferred to be made by the missionary fathers, who had long ago attempted and failed to discover this route.[15]

Even though there is a sketchy journal of the Armijo trip, there is some uncertainty about the exact route taken by his party through the eastern Mojave Desert. It seems likely they did not follow the route taken by the Salt Lake-to-Los Angeles trail along the Amargosa River exactly. Probably Armijo's general course was somewhat south of that route and may have been closer to the Kingston Springs cutoff. It does seem likely, however, that he was at Bitter Springs—which he describes as "little spring of the lava beds, or badlands"—on January 19, 1830. The next day he was a "Day's journey without water" which was a normal circumstance on the stretch of road between Bitter Springs and the Mojave River. On January 21 he reported his arrival at the Mojave River as follows: "At the arroyo of the Hayatas, at the end of which comes in the trail from Moqui, traveled by the Moquis with the object of trading shells with the said Hayatas."

If he struck the Mojave River at the point where the later Salt Lake Trail came in, he was about nine miles upstream from the site of the present Camp Cady Ranch.

Of the greatest importance to us is his recognition of the existence of a trail coming from "Moqui." "Moqui" refers to the Hopi Pueblos of northeastern Arizona and "Hayatas" may refer to the Mohaves. This entry suggests that the Mohave Trail was in existence and being used in Armijo's time, as it was more than fifty years earlier when Garces saw it. And, perhaps even more interesting, it shows that Armijo (or the people with him) had knowledge of the tribes and trails across northern Arizona. He knew what tribes were involved and he seemed to know what business was carried on between them.

Armijo arrived in California in January of 1830 and after delaying for a few weeks returned to New Mexico, arriving there in April. For the next eighteen years trains like Armijo's made the trip from New Mexico to California and back. Woolen products from New Mexico were carried over twelve hundred miles to California and were traded for horses and mules which were driven back to New Mexico. The route also became a highway for adventurers who used it to raid the rich horse and cattle ranches of California. Also, the trail served as something of an emigrant route. A number of trappers from New Mexico traveled this route to California on ventures that normally would have resulted in their returning to New Mexico—but many stayed and became leading citizens in Mexican and early American California. The Old Spanish Trail was largely a pack trail but it appears that a few wagons may have made the trip.

It is of the greatest importance to us to recognize that some of the traffic that traveled over the Old Spanish Trail to California used the ancient Indian Trail across the eastern Mojave Desert. The Wolfskill/Yount expedition over the trail in 1830-1831—one of the earliest and best known parties to use the Old Spanish Trail—made use of the Mohave Indian Trail connection.

There were at least seventeen men in the Wolfskill/Yount party—perhaps a few more. Following is a description of the route, taken from an 1866 newspaper account at the time of Wolfskill's death:

> Last of Sept., 1830, the party, with Mr. Wolfskill at its head, left Taos for this then far off Territory of California. They came by a route farther north than that usually adopted by the Spaniards in traveling between California and New Mexico—their object being to find beaver. They struck

> the Colorado just below the mouth of the Dolores, at the head of the "Great Cañon," where they crossed; entering the Great American Basin, striking the Sevier; thence southward to the Rio Virgin, which they followed down to the Colorado; thence descending the Colorado to the Mojave [Indian Villages]; where they hoped to obtain some provisions of which they were in want, and also to find Beaver. From there they took across to the sink of the Mojave river [Soda Lake presumably], through the Cajon Pass to San Bernardino, and finally to Los Angeles, where they arrived in February, 1831.[16]

Yount and Wolfskill both settled in California becoming important pioneers–Yount in Napa County and Wolfskill in Los Angeles County. Behind Yount and Wolfskill–throughout the 1830s and most of the 40s–a steady trickle of traffic inched its way over the Old Spanish Trail between New Mexico and California.

The records of the comings and goings of these parties are brief. Sometimes the scant information merely establishes that a caravan had arrived from New Mexico and that it perhaps departed at some time later. Or, there might be a reminiscence of a trip recalled many years after the fact. Most of the data do not provide descriptions of the trails sufficiently detailed to establish exact routes used across the eastern Mojave Desert. The data do establish, however, that more than one path was used. Apparently there was a route (similar to the later Salt Lake Trail) that followed the Amargosa River route and its variants. It also appears there was a second route (similar to that used by Wolfskill and Yount) that followed the ancient Mohave Indian Trail across the desert to the Colorado River.

In April of 1844 Lt. John Charles Fremont arrived on the Mojave River with his famous exploring party then returning to the east. In 1843 Fremont had surveyed the Oregon Trail to the Columbia River. During the winter of 1843-1844 he had crossed the Sierras with a small party of about twenty men entering Mexican California. He had refitted at Sutter's fort on the Sacramento River and then headed south through California's Central Valley seeking a passage through the mountains. Crossing through the Tehachapi Pass he searched for the Old Spanish Trail which he struck near the Cajon Pass.

He followed the Old Spanish Trail to the Mojave River and on down that stream to the site of future Camp Cady. While camped at Cady he was joined by two Mexicans (Andreas Fuentes and Pablo Hernandez). Fuentes and Hernandez were the survivors of a small party of Mexicans that had been attacked by Indians at Resting Springs on the main trail to Las Vegas near the Amargosa River. Fremont commented: "I received them kindly, taking them into my own mess, and promised them such aid as circumstances might put it in my power to give." The next day's journal entry begins with the following: "We

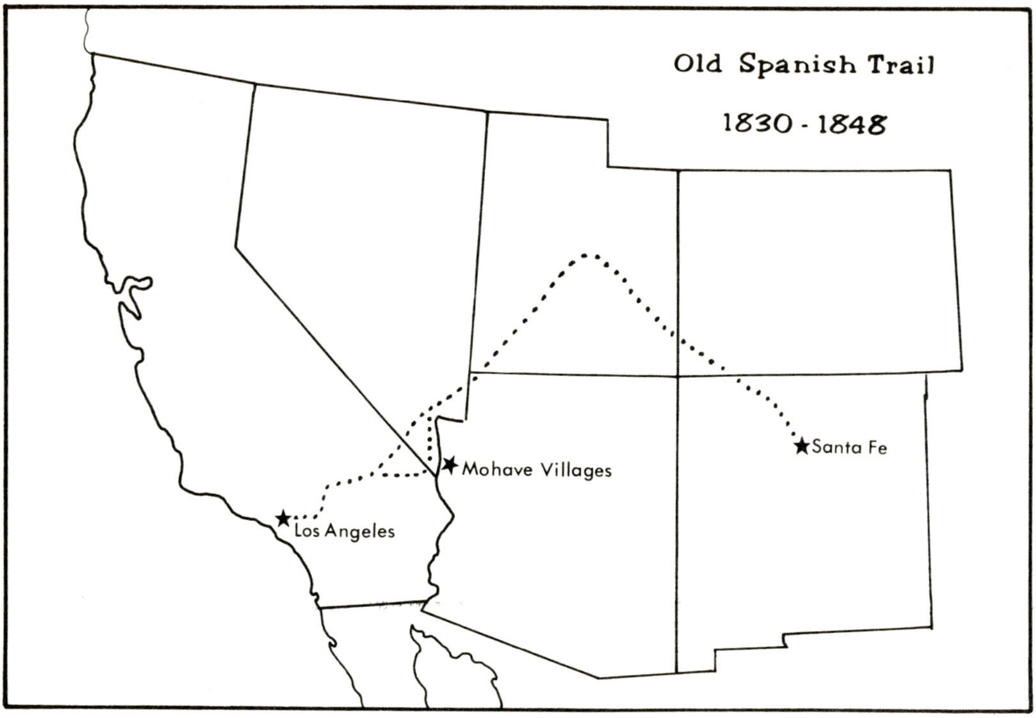
Old Spanish Trail
1830 - 1848
Santa Fe
Mohave Villages
Los Angeles

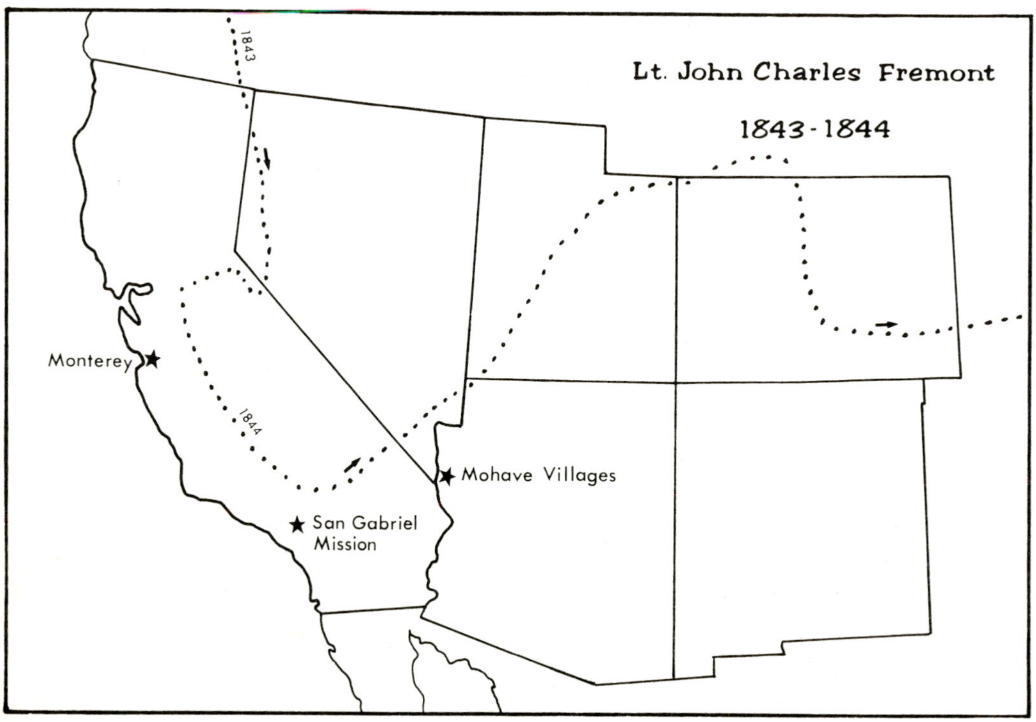
Lt. John Charles Fremont
1843 - 1844
1843
Monterey
1844
Mohave Villages
San Gabriel Mission

John Charles Fremont, From the book, *Memoir of the Life and Public Services of John Charles Fremont* by John Bigelow.

Kit Carson, From the book, *Memoirs of my Life,* by John Charles Fremont.

left the river abruptly, and, turning to the north, regained in a few miles the main trail, (which had left the river sooner than ourselves.)"[17]

Inasmuch as Fremont had passed up the usual turnoff for the Amargosa River Route (the point that later became known as "Forks of the Road") it seems likely that he did not intend to go by that exact route. Kit Carson was with Fremont and since Carson had been over the Mohave Indian Trail as early as 1829-1830 he could have informed Fremont that the Colorado River could be reached in this direction.

While camped at Cady, Fremont also made the following comment which is considered significant: "The caravans sometimes continue below, to the end of the river, from which there is a very long *jornada* of perhaps sixty miles, without water." The "end of the river" must refer to Soda Lake. The "*jornada* of perhaps sixty miles" may suggest a northern or northeastern course from Soda Lake since a course east from that point would encounter water in less than sixty miles. A northerly course from Soda Lake would bring the traveler back on the main branch of the Old Spanish Trail and this seems to me like a reasonable interpretation of Fremont's remarks.

Whether Fremont intended to continue downriver from Camp Cady or not we shall probably never know. He tells us "We left the River abruptly, and, turning to the north, regained in a few miles the main trail." He does not say definitely that this abrupt departure from the river constituted a change in plans occasioned by the difficulties of the destitute Mexicans, but that conclusion could easily be inferred from his statements. In any case, Fremont's party was soon back on the "main trail" in company with the Mexicans.

Upon arriving at Bitter Springs it was discovered that horses abandoned there by Fuentes and Hernandez had been driven off by Indians. Fremont's party waited there while Kit Carson and another trapper named Alexis Godey took the trail of the Indians. The two men overtook the depredators and collected "coup" and many of the horses before returning to join Fremont at Bitter Springs. The story of this episode as written in Fremont's report contributed a great deal to Carson's rise to national fame. Continuing on to the *Archilette*—or Resting Springs—Fremont's party discovered the bodies of the two men remaining in the Mexican camp and supposed the two women that had belonged to the party had been carried off into captivity. However, it was later learned that they, too, had been murdered. (An extract from Fremont's report, covering the period when he was on the Mojave River, is given in the appendix to this book.)

EARLY AMERICAN EXPLORATIONS

SITGREAVES, AUBRY, WILLIAMSON, & WHIPPLE

Old Spanish Trail traffic ended with the beginning of the American era following the Mexican War. The last caravans are thought to have used the route in about 1848.

Commercial factors are frequently mentioned in explaining the decline in use of the trail; but I am inclined to think, however, that commercial considerations were not so importantly involved. I think it was that "damned Yankee spirit" that did it. A good red-blooded American of the day would be offended at the suggestion that he should travel twelve hundred miles when the trip could be made just as easily (from a geographical point of view) in eight hundred miles. Bands of hostile Indians might be feared and their existence could be the cause of delays and extensive preparations–but they would not turn the course of commerce or emigration or America's "Manifest Destiny."

Two important wagon roads were blazed to southern California shortly before 1850. The first, marked out during the Mexican War by Philip St. George Cooke and his famous Mormon Battalion, traversed southern Arizona and crossed the Colorado River at what became the Yuma Crossing. This road was frequently called the Southern Emigrant Route or, later, the Butterfield Overland Mail Route. The second important early wagon road to southern California was the Salt Lake Trail. This wagon road connected Salt Lake City and Los Angeles and used part of the Old Spanish Trail. It was firmly established as a wagon road by California-bound emigrants during the Gold Rush in the winter of 1849-1850. It would yet be ten years (until 1859) before a wagon road was established over the ancient Mohave Indian Trail to the Colorado River.

The first interest in the 35th Parallel Route as a possible route to connect the settlements on the Rio Grande with southern California was originated by U. S. Army authorities in New Mexico. In 1849 Bvt. Lt. Col. John M. Washington, Chief of the Ninth Military Department and at that time military Governor of New Mexico, conducted an extensive campaign against the Navajos situated in what is now northwestern New Mexico and northeastern Arizona. His purpose in waging this war was to provide protection for the citizens of the New Mexico settlements. At the same time he was attacking one of the tribes that had for so many years blocked a direct route to California.

Perhaps even more important to us, Lt. Col. Washington had with his command James H. Simpson, a young lieutenant of the Topo-

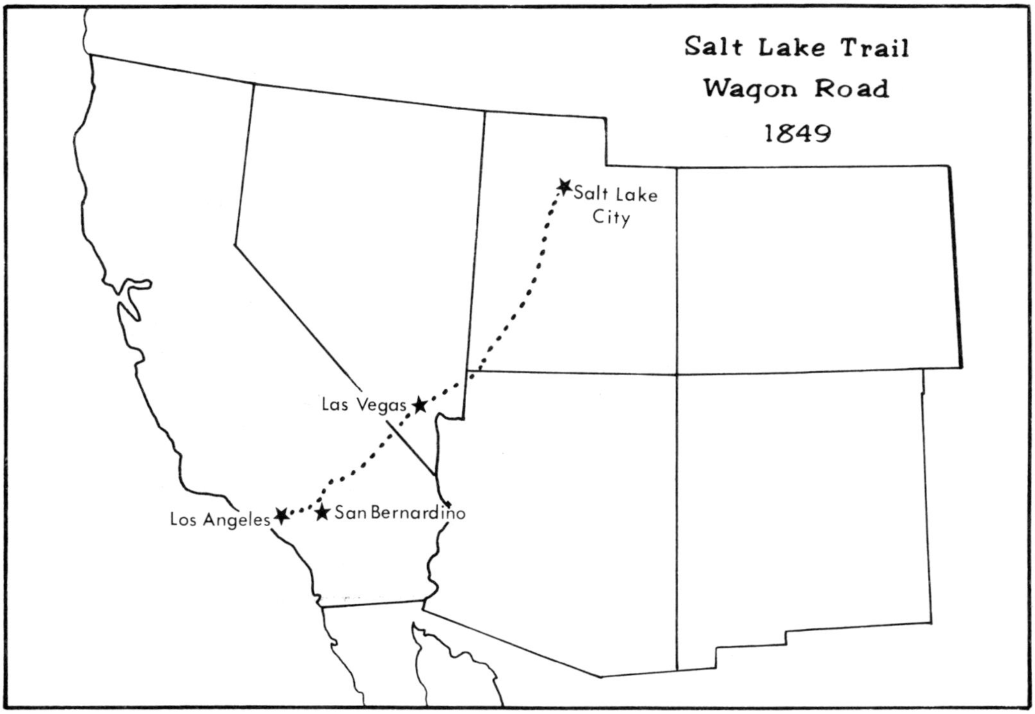
Salt Lake Trail
Wagon Road
1849
Salt Lake City
Las Vegas
Los Angeles
San Bernardino

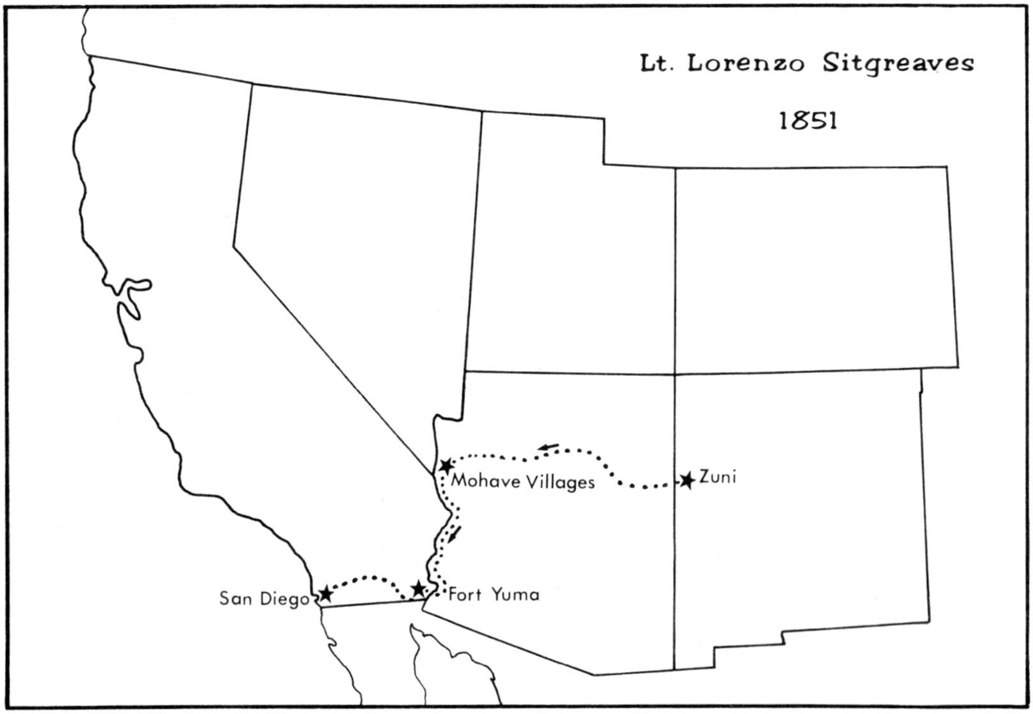
Lt. Lorenzo Sitgreaves
1851
Mohave Villages
Zuni
San Diego
Fort Yuma

graphical Engineers. Simpson produced an extensive report covering the geography and inhabitants of this little-known country. In his report he provided the stimulus that would result in the first serious exploration of the 35th Parallel Route across Arizona. The following statements are taken from the concluding remarks in Simpson's report:

> Before concluding my journal, I think it proper to bring to the notice of the department the expediency of having the country examined west of the Pueblo of Zuñi, for the ascertainment of a wagon route from the former point to the Pueblo de los Angeles, or, failing in this, to San Diego. ... ...
>
> By reference to the map, it will readily be seen that a route from Santa Fe to Pueblo de los Angeles, in the direction suggested, running as it would intermediate between the southern detour of Cooke's route and the northern detour of the "Spanish trail" route, or, in other words, as direct as possible, would not only be shorter by probably as much as three hundred miles than either of these routes, but, passing by the pueblos of Laguna and Zuñi, and possibly of the Moquis, situated still further westward, would furnish supplies of subsistence and repairs of outfit for certainly the first two hundred, if not three hundred, miles of the way–*desiderata* certainly not to be disregarded.[18]

Simpson described the possibilities of the 35th Parallel Route quite accurately, although his geographical information of points to the west of where he had actually been was quite limited and defective. He accurately guessed at the important role the Zuni pueblo would play in the development of a route of travel along the 35th parallel.

In response to Simpson's recommendations, in 1851 an expedition under Lt. Lorenzo Sitgreaves, Topographical Engineers, was sent into the country west of Zuni to the Colorado River. The instructions given to the party included the following:

> The river Zuñi is represented on good authority to empty into the Colorado, and it has been partially explored by Lieutenant Simpson to the pueblo of Zuñi. You will therefore go to that place, which will be, in fact, the commencing point of your exploring labors. From the pueblo of Zuñi you will pursue the Zuñi to its junction with the Colorado, determining its course and character, particularly in reference to its navigable properties, and to the character of its adjacent land and productions. The junction of the Zuñi and Colorado will be accurately determined. You will then pursue the Colorado to its junction with the Gulf of California, taking those observations which will enable you accurately to delineate its course.[19]

Sitgreaves organized his party at Santa Fe and it consisted of: "Lieutenant J. G. Parke, Topographical Engineers; S. W. Woodhouse, M.D., physician and naturalist; Mr. R. H. Kern, draughtsman; Mr. Antoine Leroux, guide; five Americans and ten Mexicans as packers and *arrieros*." Additionally, he had an escort of thirty-two men of Company "B" 2nd Artillery commanded by Capt. Henry L. Kendrick.

The instructions given to Sitgreaves reveal the lack of accurate geographical information regarding northern Arizona that existed at the time. If he followed these instructions to the letter (a thing that would have been virtually impossible) he would have continued down the Zuni River to the Little Colorado River and down that stream to the main Colorado. Then he would have traveled down through the full length of the Grand Canyon! Needless to say, the objective of locating the junction of the Little Colorado with the main stream was abandoned early as Sitgreaves approached the broken canyon country. In fact, it appears that he had known before he arrived at the falls of the Little Colorado that he could not follow the river much farther because of the canyoned and broken country.

Upon the advice of his guide—and citing the condition of his pack animals and rapidly dwindling supplies also—Sitgreaves determined to leave the Little Colorado near its lower canyon and turn west intending to strike the main Colorado below the Grand Canyon. He passed to the north of the San Francisco Mountains. Near Bill Williams Mountain he struck what he supposed were the headwaters of the Bill Williams River. However, the streams were running south instead of west so he was forced to leave this reliable water source in order to pursue his western course.

From Bill Williams Mountain to the Colorado River, a distance of nearly two hundred miles, the country was mostly desert. There were watering places in this country but their location was not known to anyone in the Sitgreaves party. Consequently, the party suffered greatly on this trek. Also, the Indians were more or less hostile and at one point they succeeded in wounding the expedition's guide, Antoine Leroux, so severely that he was considered unable to function for the rest of the trip.

Finally, on November 5, 1851, the expedition crossed the Black Mountains (probably near what is now known as Union Pass) and struck the Colorado River in the vicinity of the present Bullhead City, Arizona. This put the expedition at the northern end of the Mohave Villages. Sitgreaves provided the following description of their first glimpse of the long-sought Rio Colorado of the West: "We passed through [the mountain pass], however, unmolested, and were at length cheered by the view of the Colorado, winding far below through a broad valley, its course for many miles being apparent from the large trees upon its banks. The smoke of numerous fires in the valley gave evidence of a large Indian population, and the sight brought a spontaneous cheer from the men, who believed that this was to be the end of their privations and of the labors and anxieties of the journey."

Sitgreaves was to be disappointed in his hopes for relief and assistance from the Mohaves. He found them to be suspicious at first and later openly hostile. Whether the Indians or the explorers were responsible for this is difficult to determine with available data. An extract from the Sitgreaves report, covering the period he was among the Mohaves, is given in the appendix in this book.

The expedition continued on down the Colorado, traveling on the east bank. On November 16 they came upon a large settlement of Yuma Indians. At first the Indians were friendly but later they killed a soldier of the escort who had lagged to the rear. This was followed by a general attack by the Indians during which the expedition was able to hold its own and killed at least four of the attackers.

The expedition finally arrived at Fort Yuma on November 30, 1851. They were in desperate condition and had been living on mule meat for several days.

The Sitgreaves expedition had traversed a country previously unknown to the American scientific community. But, unfortunately, the expedition did not contribute much accurate geographical information about this country. It was a hurried reconnaissance. The explorers were preoccupied with the business of finding water and being watchful of the weak but hostile and persistent Indians along the route.

So far as is known, Sitgreaves did not state whether or not he thought a wagon road existed across northern Arizona as suggested by Simpson. Conclusions are missing from his brief report and the reader can infer only general impressions from his day-to-day journal entries. The tone of these comments is pessimistic. The impression of a waterless country, infested with never-ending bands of hostile Indians, is conveyed—certainly not attractive-sounding as a possible wagon route. But Sitgreaves refrained from drawing conclusions.

In the late 1840s and early 1850s the dominant topic on the national scene in America was the question concerning a railroad that would reach from the Atlantic to the Pacific. The feasibility of such a route had been discussed for years but it had barely gotten beyond the talking stage. In the minds of many the project seemed totally impossible—actually beyond the capability and technology of man. But, in view of the huge land acquisitions that resulted from the Mexican War and the subsequent discovery of gold in California and its admission as a State, a Pacific railroad seemed not only *practicable* but *absolutely necessary* to protect the newly won empire and to guarantee the cohesiveness of the far-flung parts of the Union.

The Pacific Railroad was on the minds of men everywhere. This subject dominated the press, the thoughts and actions of Congress, and even the discussions around evening campfires of the multitude of emigrant and freight trains plying the wagon roads to the Promised Land.

Francois-X. Aubry

Courtesy, National Library of Canada, Ottawa

Inevitably, these discussions led to the question of what route the road should take. Every route had its champions. Some individuals were inspired to speak out because they genuinely believed that their favorite route was superior to the others. More frequently, however, ulterior motives were just under the surface. Every town along the Mississippi River thought it should be the terminal point for the great road to the Pacific. Immense commercial advantages would, of course, result. Senators and Congressmen defended their sections and special interests. Southern men thought it should be a southern route–northern men thought it should be a northern route. Anyone with knowledge of a specific route would be called to testify and provide information. Underlying it all was a great dearth of accurate geographical information about the West. No one knew for sure whether there *was* a suitable route over which a railroad could run to the Pacific.

Private citizens became embroiled in the controversy–they were interested in it–it was a fascinating and absorbing subject. In the years 1853-1854 the U. S. Army Corps of Topographical Engineers was directed by Congress to conduct a number of detailed and extensive surveys in the West to "ascertain the most practicable and economical route for a railroad from the Mississippi River to the Pacific Ocean." They were naive enough to believe that, if enough scientific data could be compiled, the weight of those data would indicate where the railroad should go; and, hence, sectional controversy would be ended. While the government was making up its mind to conduct this reconnaissance (government moved about as rapidly in the 1850s as it does now), a private citizen, Francois-Xavier Aubry, became inspired to explore the 35th Parallel Route with his own resources. It is difficult to find an explanation for Aubry's decision to become involved in this expensive and dangerous undertaking–except that he was interested in the controversy and he was by nature a man of direct and decisive action.

Aubry was a well-known Santa Fe merchant. He had gained national recognition from the speed with which he was able to make trips from Independence, Missouri, to New Mexico over the Santa Fe Trail. He had earned the sobriquet *Skimmer of the Plains*.

In 1852-1853 Aubry changed his field of operations by driving a herd of sheep and taking some wagons to California from New Mexico. He followed the southern route, crossing the Colorado River near Yuma. For the return trip to New Mexico he explored the direct route along the 35th parallel.

Aubry, with a party of eighteen men, crossed the Sierra Nevada at Tejon Pass and headed east on July 11, 1853. They struck the Colorado River on July 22. His precise route across the Mojave Desert and the exact point at which he struck the Colorado are somewhat in doubt. Aubry did leave brief journals of his trips but they are not sufficiently detailed to settle definitely the question of route (these journals have been published several times and consequently are not repeated here).[20]

My feeling about Aubry's route is as follows. It would appear he struck the Mojave River a little below the present Victorville. He followed the usual trail to the site of the future Camp Cady. Here he left the Mojave River Trail and headed northeast. From there to the vicinity of Las Vegas he presumably followed a course approximating the present Interstate Highway 15. The Clark Mountains could have been crossed either through Mountain Pass or the more northern Keany Pass. The final day's journal entry before reaching the Colorado River read as follows: "Traveled twenty miles east southeast, most of the distance through a little canyon, where we found good grass, water, and cane in abundance, and struck the Great Colorado of the West." Some writers have speculated that Aubry reached the Colorado by way of El Dorado Cañon. The description sounds more like Las Vegas Wash to me—but either view is tenable.

From the Colorado River Aubry passed on across northern Arizona to the north of the San Francisco Mountains, as Sitgreaves had done. His small party was repeatedly attacked by Indians in Arizona and Aubry and many of his men were wounded. They arrived in destitute condition at the pueblo of Zuni on September 6, 1853. On September 10 he was back in Albuquerque and a few days later arrived in Santa Fe.

Aubry's diary and his observations about the 35th Parallel Route were published in a Santa Fe newspaper and copied in other papers of the day. The editor in Santa Fe commented as follows on Aubry's achievement:

> We have the satisfaction of laying before the public Mr. Aubry's notes of a trip from California to New Mexico by a route hitherto but little known. In fact the route pursued by Mr. Aubry has never before been explored by any one. The belief which has long existed in the public mind of New Mexico, in regard to the existence of an excellent route for a wagon or railroad from the central part of the Territory to California, has received ample confirmation from this exploring expedition of Mr. Aubry. From this enterprise of noble daring, the most interesting and useful information is furnished to the world, on a subject, which at this time, engrosses the attention of statesmen and capitalists to a degree perhaps unknown in any former period of our history. This information too, although unofficial, is not the less authentic or the less worthy of credence on that account. To those who know Mr. Aubry it would be superfluous for us to say a word in behalf of his veracity and good judgment. His friends all know him to be a man absolutely without a parallel in physical qualities, and unsurpassed in all the noble traits of human character. We confidently give assurance to our readers that the journal of this gentleman may be relied upon to the letter,

and that we have reason to believe that the modesty which always attends real worth, has caused him to under-value the usefulness of his labors, and the accuracy of his judgment.[21]

In October of 1853–just a few weeks after arriving in New Mexico–Aubry was once again on his way to California over the southern route with a drove of sheep. He sojourned in California for several months and by July of 1854 was ready to return to New Mexico. With a well-armed party of sixty men he was prepared to face the 35th Parallel Route and its hostile Indians a second time. His stated purpose was "to locate a wagon road from this valley to Albuquerque on the north side of the Gila, in the thirty-fifth parallel of latitude or as near it as practicable." He followed nearly the same route as in the previous year but this time took a wagon to carry a boat for crossing the Colorado River and to demonstrate the practicability of the route for wagon travel. The wagon was taken through to New Mexico.[22]

Aubry's second trip across Arizona was less eventful than it had been the previous year. He was eager for revenge against the Indians who had caused him so much trouble the previous year, but the wily children of the mountains would not risk an engagement with Aubry's sixty well-armed men. Aubry stated that this second expedition was outfitted at a cost of about $15,000–a considerable sum in those days.

By mid-August Aubry was back in New Mexico. Aubry might well have figured importantly in the subsequent history of the 35th Parallel Route except his life was to be cut short. Shortly after arriving in Santa Fe from his second trek over the route he was killed by Richard H. Weightman in a row growing out of the railroad route controversy on August 18, 1854.[23] Aubry was in his thirtieth year.

Before leaving the subject of Aubry's explorations along the 35th Parallel Route, a question should be raised why he was motivated to select the specific route that he chose to cross the eastern Mojave Desert. Why did he go so far out of his way and cross the Colorado River at a point so far north? It would have been much easier to travel directly east from the Mojave River over the old Indian Trail and strike the river at the Mohave Villages. Aubry implies in his diary that he did not proceed directly east from the Mojave River because of sand. This seems like an inadequate explanation. There was a road through this country that avoided most of the sand, with water at convenient distances, and Aubry could have found out about it. Also, in the 1853 trip, he had only a pack train and could have negotiated the sand easily.

I am inclined to think it quite likely that Aubry did not want to face the Mohaves and preferred not to admit it. I think this may be the reason he went so far out of his way to cross the river. I should point

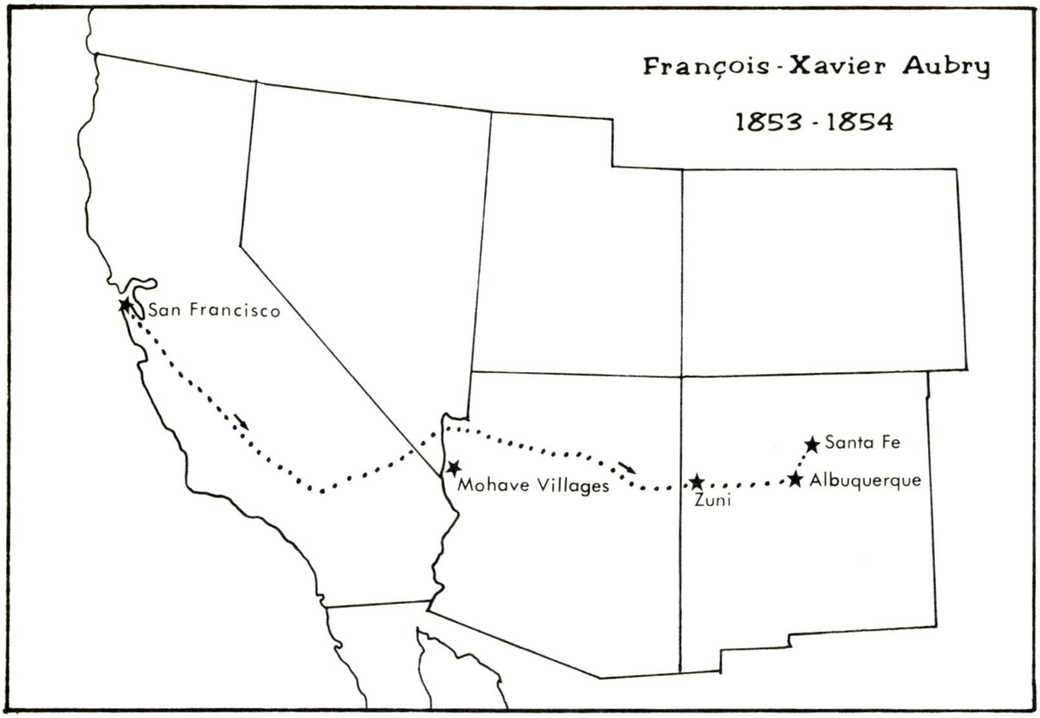
François-Xavier Aubry
1853 - 1854
San Francisco
Mohave Villages
Zuni
Albuquerque
Santa Fe

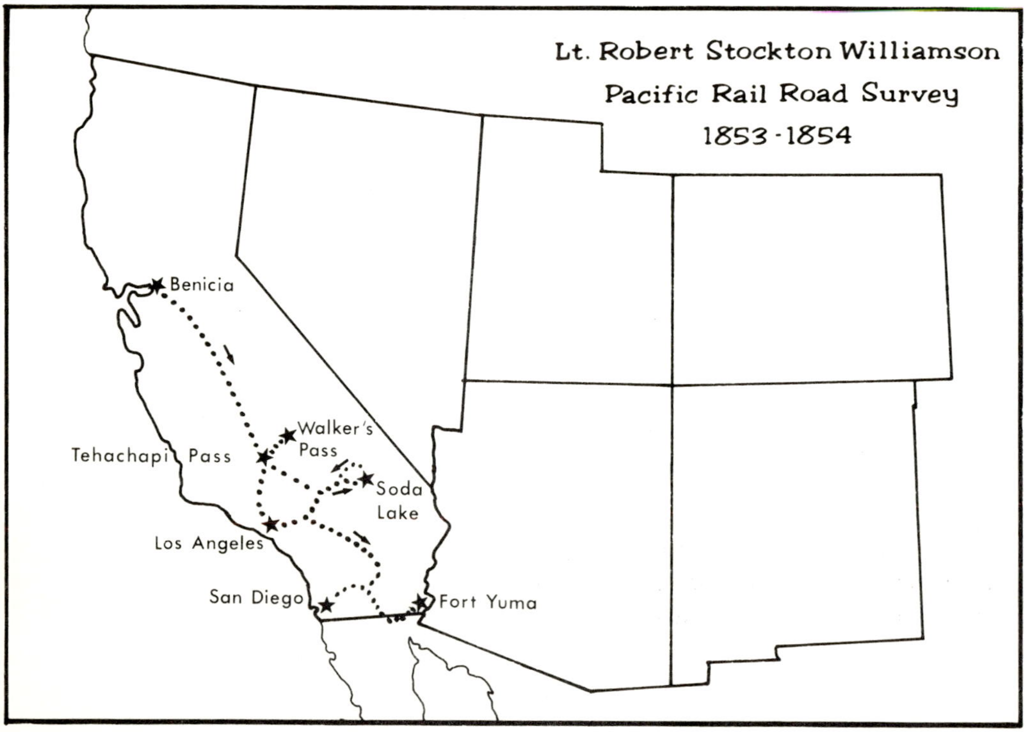
Lt. Robert Stockton Williamson
Pacific Rail Road Survey
1853 - 1854
Benicia
Walker's Pass
Tehachapi Pass
Soda Lake
Los Angeles
San Diego
Fort Yuma

out that this is speculation on my part—Aubry barely mentions the Mohaves at all. In fact, his silence on this important point may be significant.

Having a reluctance to face the Mohaves with only eighteen men represents good judgment. Depending upon what frame of mind he found them in (and how politic he was in his dealings with them) he might well have gotten his entire party massacred if he had decided to cross at the Mohave Villages. Aubry appears to have made greater efforts to go through direct to the Mohave Villages in 1854 than he did in 1853. And certainly a man of his experience on the frontier would have been safe enough at the Mohave Villages with sixty well-armed men.

I do not raise this point to cast any negative light on Aubry's ability or character—he is well-established in the annals of western history as a truly remarkable person of the highest character. I am of the opinion that Aubry did take his more northern route because of the powerful Mohaves; and I think that observation has a place in this narrative as another instance where Mohave presence caused an indirect route to be taken.

Aubry's private investigative expeditions of 1853 and 1854 have few parallels in the history of western exploration. Without a doubt his exploits would be better known if his life had not ended so abruptly and ingloriously at the end of his second expedition. Another factor obscuring Aubry's achievements is the great government surveys that were conducted in 1853-1854 by the Army Corps of Topographical Engineers under the auspices of the War Department. These surveys were going on simultaneously with Aubry's work and, in fact, both of the railroad surveys of interest to this paper took place between Aubry's two treks.

The Military Appropriations Bill—including the funds for the Pacific Railroad Surveys—was approved by Congress in March of 1853. The work of the surveys was to be made available to Congress within ten months. Needless to say, the War Department was unable to comply completely with this unrealistic deadline.

By the time it had ended, four major east-west lines were examined by the engineers plus a detailed exploration of passes through the Sierra Nevada Mountains. Of the four major lines one was between the 47th and 49th parallels, one between the 38th and 39th parallels, one along the 35th parallel, and one along the 32nd parallel. The expeditions of particular interest to this narrative are the one under Lt. Robert S. Williamson, who surveyed passes through the mountains of southern California, and the one along the 35th parallel under Lt. Amiel Weeks Whipple.

Williamson's was the first survey to touch the eastern Mojave

Desert. He left Benicia with his expedition to commence his labors in Tulare Valley and to examine some of the southern passes through the Sierra Nevada on July 10, 1853. He probably did not know that on that same date Aubry and his little band of eighteen determined men were gathered at Tejon Pass ready to plunge into the unknown desert wilderness on their first trek to New Mexico over the 35th Parallel Route.

Of the two surveys that touch upon the subject of the 35th Parallel Route (i.e., Williamson and Whipple) Williamson's work was by far the more competently done. He made important contributions to the understanding of the railroad geography of southern California.

Williamson spent the summer and early fall of 1853 examining the Tulare Valley and passes through the southern Sierras. Considerable time and energy were expended in determining definitely that the famous Walker's Pass (which previously had been one of the prime attractions as a passageway through the mountains for a Central Route) was not practicable for a railroad. Then, he journeyed slightly south and located two passes–the Tehachapi and the Cañada de las Uvas–that were practicable for railroad use. His instructions now provided that he should attempt to locate a route across the desert that might connect the 32nd Parallel Route (Cooke's Wagon Road or the southern Emigrant Route) or the 35th Parallel Route to these newly discovered passes through the Sierras. This brings Williamson into the eastern Mojave Desert and hence within the scope of this paper.

On many early maps the Mojave River is shown as winding across the eastern Mojave Desert and finally entering the Colorado River near the Mohave Villages. Railroad promoters interested in the 35th Parallel Route had long viewed this mythical course of the Mojave River as an easy route for the transcontinental line from the Colorado River westward to the Sierras. There was some question about whether the Mojave River actually reached the Colorado–as well there should have been since it actually does not–and therefore this was an important point to settle. Mountain men had known for a long time that the Mojave did not empty into the Colorado River but the scientific community and pro-35th parallel railroad promoters were not convinced.

Williamson grasped the importance of this point and expended the necessary time and energy to settle the issue. In the course of doing so he generated an interesting early written account of the lower Mojave River and Soda Lake country. For this reason that part of his report is quoted in full in the appendix to this book.[24]

On July 14, 1853–just a few days after Aubry left Tejon Pass on his first exploration and shortly after Lieutenant Williamson left Benicia on his exploration–Lt. Amiel Weeks Whipple commenced the formal survey of the 35th Parallel Route at Fort Smith, Arkansas. He

Robert Stockton Williamson
Courtesy, United States Military Academy Archives, West Point

would measure off more than eighteen hundred miles with his odometer before his expedition reached Los Angeles and the Pacific Coast in March of the following year.

The route across Oklahoma and eastern New Mexico was fairly well-known. There is a real question about whether so much time was justified in conducting a reconnaissance of a distance of more than eight hundred miles that was already so well-known. It was not until October 5, 1853, that the expedition arrived in Albuquerque—several weeks after Aubry's party arrived from California. During the next month Whipple made preparations for the trek into the relatively unknown country west of Zuni. He obtained information from Aubry and some of his men and he managed to retain Antoine Leroux as a guide. Leroux had been guide for the Sitgreaves expedition in 1851.

Whipple finally left Albuquerque, heading west on November 10, 1853. His train was quite large. There were about fifteen professional people acting in such capacities as surgeon, botanist, astronomer, secretary, surveyor, geologist, topographer, artist, and engineer. Then, there were three army officers in addition to Lieutenant Whipple and his assistant Lt. Joseph Christmas Ives. Two of these additional officers (Lt. John M. Jones, 7th Infantry, and Lt. John C. Tidball, 2nd Artillery) were assigned to the escort. The third, Lt. David S. Stanley, 2nd Dragoons, was the expedition quartermaster and commissary. Then, when a number of men from Fort Defiance had joined the expedition, the escort consisted of over fifty enlisted men. These troops were from companies of the 3rd and 7th Infantry and the 2nd Artillery. Also, there was a large train of wagons, a large drove of mules, and the usual assortment of teamsters, packers, and herders. There must have been well over one hundred men and at least that many animals.

The route as far as Zuni, at least, was well-known. After leaving that place the party had the feeling of entering a *terra incognita.* Lieutenant Stanley confided to his journal: "here ends our traveling over a path that has ever been passed over by the white men before. To-morrow we move into an untraveled country, and what is still more discouraging, we learn that the road is tough." Of course Aubry and his party had passed through this same country just about two months earlier—and many other white men had been there before Aubry. But at least it shows that the expedition was inspired with a feeling of adventure when they cut loose from all lines of communication and headed west at Zuni.

So far as adding to existing knowledge about railroad geography, there were probably two main objectives Whipple's expedition could have accomplished, but it failed at both. One was to unravel the country on a direct line over the 35th parallel from the San Francisco

Mountains to the Colorado River and the other was to settle the point as to whether or not the Mojave River ran into the Colorado.

It was already known that a route could be found to the San Francisco Mountains from Zuni—although earlier travelers (both Sitgreaves and Aubry) had gone around the north side of the mountains while Whipple did discover the pass (now used by Interstate Highway 40 and the Santa Fe Railroad) to the south of these mountains. But that was not the big problem. From the San Francisco Mountains directly west to the Colorado River the country was more broken and at lower elevation became a desert. A route was needed through this region. Sitgreaves and Aubry both had found a way through it but Sitgreaves had not surveyed the route thoroughly and Aubry had just recently completed his reconnaissance. It is to be imagined that Whipple would not want to make extensive use of Aubry's findings, anyway, for fear of revealing that he had been preempted by a private citizen.

Whipple spent many weeks trying to find a suitable passage for his large train along the 35th parallel through the desert country, but without success. He gained information from some Indians who told him there were only three small springs between him and the Colorado River on the desert route, and this was not enough for the more than one hundred and fifty miles of unknown country. At this time he was on the headwaters of the Bill Williams Fork, which was known to empty into the Colorado. However, the stream was running southwest instead of west. Finally, on January 29, 1854, Whipple yielded to the temptation and committed himself to follow the Bill Williams Fork to the Colorado. Once committed to this course there was no turning back. His hopes sank as the river drew him farther and farther south over a route that would be totally impractical for a railroad. Ultimately, when he struck the Colorado at the mouth of the Bill Williams Fork, he was nearly sixty miles below the 35th parallel.

As the expedition toiled down the sandy riverbed, the mule teams began to give out. One by one wagons were left behind until, when they finally reached the Colorado on February 20, 1854, only six wagons and a small cart remained with the expedition. At this time the six wagons were abandoned and only the little cart was taken on through to California to carry the expedition's instruments.[25] Whipple now ascended the Colorado to a point opposite the vicinity of the present Needles, California, where he crossed the river to the west side on February 27, 1854.

Whipple and his party were, of course, among the Mohaves from the time of arriving on the Colorado near Bill Williams Fork until they left the valley of the Colorado. For several days they were encamped in the immediate vicinity of the Mohave Villages. Whipple found the

Amiel Weeks Whipple
Courtesy, National Archives

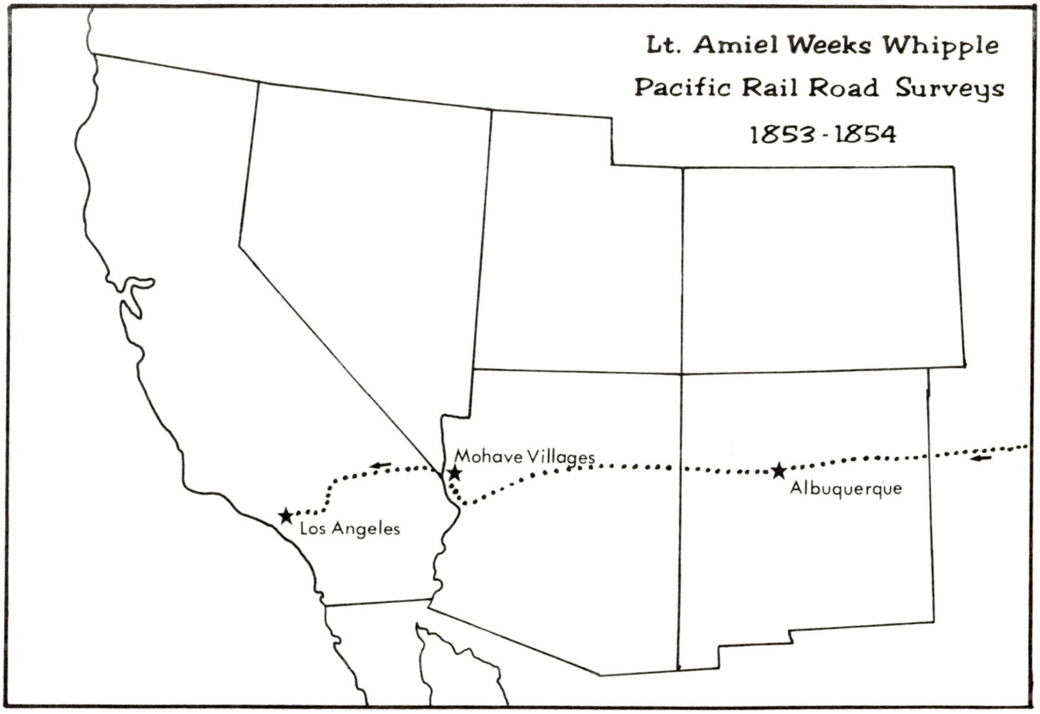

Mohaves a little difficult to manage at times—but, generally speaking, he got on very well with them. Whether this relationship was a result of good management on his part or the large size of his party is difficult to determine. I am inclined to believe it was a mixture of both. Whipple seemed to know how to handle the great hordes of Indians that came to visit and trade—this is the point at which many sojourners in Mohave Country made the errors in judgment that led to hostilities.

Whipple negotiated successfully with the Mohaves to obtain guides to lead him across the eastern Mojave Desert. His discussion of these negotiations is quoted immediately following for the information it provides on the roads across the desert:

> The chiefs came to-day with the interpreter to say that a national council had been held, at which they approved of the plan for opening a road to travellers through the Mojave country. They knew that upon the trail usually travelled by the Pai-Utes towards California the springs of water were small, and insufficient for our train; that on the journey our mules therefore would perish from thirst, and the expedition fail. Hence they had selected a good man, who knew the country well, and determined to send him to guide us by another route, where sufficient water and grass could be found. They wished us to report favorably to our great chief, in order that he might send many more of his people to pass this way, and bring clothing and utensils to trade for the produce of their fields. [26]

This statement tends to verify an impression already gained that there was more than one route connecting the Mohave Villages with the Mojave River. Garces certainly traveled a more southerly route than the route over which the Mohave guides led Whipple. The Garces route went through country farther south and would seem to fit the description "usually travelled by the Pai-Utes" in the above paragraph.

The route used by Whipple and his Mohave guides was very nearly that which later became the Mojave Road. His starting point was about ten or fifteen miles lower on the river than the point where the wagon road later crossed. Therefore, instead of heading directly west from the river to reach Pah-Ute Creek, Whipple's party had to round a point of mountains and then turn northwest. In that way the guide led them to Pah-Ute Creek and the next day on over Pah-Ute Hill.

From Pah-Ute Hill they traveled across Lanfair Valley to Rock Spring and from there to Marl Springs. From Marl Springs westward the road led through the Old Dad Mountain region by way of Jackass Canyon instead of passing to the north of Old Dad Mountain as did the later wagon road at times. They then crossed the southern end of Soda Lake (apparently they did not see Soda Springs) and struck the Mojave River at Cave Canyon.

While in the Soda Lake country Whipple recorded in his journal the following thoughts about the road he was passing over:

> By odometer the distance from Marl springs is thirty miles—the greatest jornada, without water, that has been encountered upon our route. But, although the mules were much worn down and wearied by the long-continued journey of many months, we have to-day lost not an animal, and neither man nor beast has suffered from thirst. For an emigrant-road it is probably the best route yet discovered across this western desert.[27]

The next day, in the same locale, Lieutenant Stanley—who seems not to have gotten on well with Whipple—recorded his thoughts on the same subject. As can be seen, the two views are exactly opposite:

> March 9th. [1854] Moved early this morning and made twelve miles. Reached the Mohave River where it is a running stream. This river, heading in the mountains of California, is a considerable stream until it reaches this sandy valley, into which it debouches from a deep, rocky canon and is immediately, totally lost. It is laid down as emptying into the Colorado in the great valley of the Mohaves, but no such connection can be traced, or was traced by our party. From the point of running water to the Colorado River the country is mountainous and sandy, with little water and grass. Unfit for waggon road or for pack mule trains. Our animals are in sad plight and we have lost many. Had we not had Indian guides, well acquainted with the country, we would most certainly have lost all our animals, if not men.[28]

This is the point at which Whipple should have settled the issue about whether the Mojave River ran into the Colorado or not. He was on the same ground Williamson had been on about four months earlier, but he did not know Williamson had been there nor whether or not Williamson had settled this important point. It would have cost him at the very most two days to make this determination—and probably one day would have been enough. Instead, he was content to hurry on through and leave the impression in his report that the point had not yet been determined—even though Williamson's preliminary results would have been available to him while his own report was being prepared in Washington. But instead of taking advantage of Williamson's information, he included the following statement in his report which, it would seem, must have been a source of considerable embarrassment to him:

> Whether a channel exists by which the Mojave river finds its way to the Colorado is a problem not yet solved. We regretted exceedingly that time for a closer examination was not at our disposal. Our trusty Indian guide, however, assured us that the dry channel of that stream passed uninterruptedly north of our route to the Colorado, and that wagons could pass through it without encountering a hill.[29]

Traveling on west, and when camped at the site of future Camp Cady, it was determined that a Mexican herder in Whipple's party had disappeared. He had been left with two tired mules the day previously down the Mojave River and had not returned. A party was sent in search of him. They found the Indians that had murdered him in camp preparing a dinner of mule meat, but the Indians managed to make good their escape. They found the bloody and arrow-riddled clothing of the herder, but his body was never found.

On March 13, 1854, Whipple announced his arrival at Forks of the Road as follows: "Having travelled about six miles we struck the Mormon road." His party then traveled the road directly to Los Angeles, making only a cursory examination of Cajon Pass along the way. They arrived in Los Angeles on March 21, 1854, and during the next two days an auction was held at which much of the property of the expedition was sold. Most of the officers of the party then departed for San Francisco where they boarded an ocean steamer for Panama and ultimately arrived in Washington, D.C., to work on their report.

It is a well-known fact of history that the Pacific Railroad Surveys did not settle the question of where the first transcontinental railroad should go.[30] In fact, it has been pointed out that they simply added fuel to the fire. The arguments became more sophisticated and technical–but the politicians did not give an inch. Perhaps the greatest achievement of the surveys was the acquisition of a tremendous amount of badly-needed scientific information about the geography, geology, and inhabitants of the American West. Also, the nation as a whole might be comforted to learn that there need be no worry about whether there was in fact a practicable route for a railroad to the Pacific. They now knew definitely that there were many such routes.

A number of extremely valuable descriptions of the country in northwestern Arizona and the eastern Mojave Desert were generated during the Whipple survey. In addition to the tome produced by Whipple himself, there is an extensive account by the artist, Heinrich Baldwin Mollhausen. Also, there are at least two diaries, those of Lieutenants Stanley and Tidball.[31]

BEALE'S WAGON ROAD

Demands for information pertaining to railroad geography were satisfied for some years after 1854. But the surveyors were not through with the eastern Mojave Desert. In an entirely different branch of the government—the General Land Office—appropriations were obtained to commence the survey of the new State of California. There were resources enough to support contracts to have the township lines surveyed throughout most of the desert country. In some cases the townships were actually subdivided.

Early contracts touching the eastern Mojave Desert were negotiated with private individuals—designated as Deputy Surveyors—in late 1854, and many others were let in 1855 and 1856. Actual field work in the areas of interest to this book was done in 1855, 1856, and 1857. During those years small parties of surveyors wandered all over the desert erecting mounds, setting stone markers, driving charred stakes along township lines, digging trenches, pits, and otherwise performing functions required by the regulations of the General Land Office.

This vast—and expensive—effort seems not to have been of lasting value. The field notes of subsequent land surveys show that in almost all instances surveyors could not relocate the stone monuments, charred posts, mounds, pits, and trenches constructed during the early years.

The surveys brought a relatively large number of people into the desert for a considerable time and resulted in the acquisition of detailed knowledge about that hitherto unknown region. The contractors for the eastern Mojave Desert included Alex P. Greene, Robert B. Hays, Alfred H. Jones, Henry Washington, and Samuel C. Wiltse. Their employees on the surveys—the chainmen, compassmen, moundsmen, and axemen—were working men and some of them would be back on the desert a few years later as prospectors or guides providing information to others who were on military missions or who were interested in mining or new roads.

The surveyors of 1855-1857 brought wagons into the eastern Mojave Desert. With the exception of the wagon Aubry used in 1854 to carry his boat to the Colorado, and Whipple's instrument cart, the surveyors were probably the first to bring wagons into the desert country south of the Salt Lake Trail. It appears quite certain that they used approximately the same route that the Mohave guides showed Whipple in 1854—except that they made use of Soda Springs and also the road ran directly east from Pah-Ute Creek to the Colorado River. In reaching the Colorado by this route they had to cross the southern tip of Nevada—but doubtless they were willing to do this to make use of such a direct and easy route to reach the river.

The surveyors naturally came in contact with the desert Indians. In the more southerly parts of the Mojave Desert around Twentynine Palms and between that point and the Colorado River, the Indians (probably Chemehuevis) were friendly and even helpful. But, to the north along the line of the Mohave Trail, the Indians (probably southern Pah-Utes) were bothersome. They raised several caches of provisions the surveyors had stored and early in 1856 they attacked one of Alex P. Greene's survey parties near Soda Lake killing two men (R. S. Redmond and Hugh McGorrey). No action was taken against the Indians for these depredations but the incidents served to strengthen the already well-established public impression that the desert Indians were hostile and deserved punishment.

The next phase of Mojave Road history begins with the first wagon road expedition along the 35th Parallel Route superintended by Edward Fitzgerald Beale. The section of road to be surveyed by Beale was from Fort Defiance across northern Arizona to the Colorado River "at a point opposite the supposed mouth of the Mohave river" (The War Department had apparently paid no attention to the important findings concerning the Mojave River presented in Williamson's railroad survey report which by this time was published in final form). Some have expressed surprise about the endpoints on Beale's survey–why from Fort Defiance to the Colorado instead of, say, Albuquerque to Los Angeles? The reason is that a wagon road already existed from Albuquerque to Fort Defiance and, as discussed above, the General Land Office had already paid to run township lines on the California stretch–certainly it would be difficult to justify an additional expenditure for basic exploration work there.

Edward Fitzgerald Beale was born in 1822 and joined the U. S. Navy in 1836 when he was only fifteen. He subsequently rose to the rank of lieutenant. He served on the Pacific Coast while in the navy and became involved in early California history. During the Mexican War he and Kit Carson were with Kearny at the Battle of San Pasqual. The two men distinguished themselves by stealing through enemy lines to bring help to Kearny's beleaguered command. He earned a reputation as a scout and frontiersman carrying messages for the federal government across the continent in the early days of American activity in California. He is said to have carried the first gold to Washington, D.C. following the discoveries that ushered in the Gold Rush. He counted men like Kit Carson and John Charles Fremont among his closest friends.

Beale resigned his Navy commission effective March 5, 1852, and for awhile engaged in the freighting business in California. Then, in December of 1852, he received an appointment as Superintendent of Indian Affairs for the State of California. Returning to California from the east coast in 1853, Beale was encouraged by Senator Thomas Hart Benton of Missouri (father-in-law of John C. Fremont) to make an exploratory trip over what Benton conceived to be the great central route to the Pacific. This Beale did in the period May through August of 1853. Benton's purpose in urging Beale to make this exploration was that this particular route had been left out in the arrangements made by the Secretary of War for the Pacific Railroad Surveys and it was the route that would logically have its eastern terminus at St. Louis—Benton's choice.

Beale was accompanied on this trip by his cousin Gwynne Harris Heap who kept a detailed journal of the trip which was published under Heap's authorship as *Central Route to the Pacific* in 1854. The route followed by Beale on this trip was through Utah and in fact close to the Old Spanish Trail along much of its course. He entered California by way of the Salt Lake Trail joining the Mojave River route at Forks of the Road.

The appointment as Superintendent of Indian Affairs terminated in 1854. In 1855 he acquired Rancho La Liebre in the Tejon country and by 1857, when he received the appointment as superintendent of the wagon road survey along the 35th parallel, he was in partnership in the cattle business with Samuel A. Bishop on this Tejon Ranch.

The Secretary of War issued detailed instructions to Beale in a letter dated April 22, 1857.[32] This letter is reproduced in the appendix to this book. These instructions show clearly one point that is frequently overlooked—the main purpose of Beale's expedition was to survey a wagon road between the stated points, and not to experiment with camels. The Congressional appropriation of $50,000 used for the expedition had nothing to do with camels. The so-called "Camel Corps" has such a broad interest-base that some writers have overlooked the main purpose of the appropriation. Beale's assignment was to survey a wagon road. As a secondary assignment he was directed to conduct a test of the camels.

The story of the War Department's importation and experimentation with camels has attracted the attention of several researchers and writers. There is no point rehashing here in detail that which has been most thoroughly done and especially since the results of this earlier work are readily available. Instead, I will provide brief background statements on how the camels happened to be at Camp Verde, Texas, where Beale picked them up in June of 1857, and I suggest the interested reader consult one of the thorough treatments of the camel subject for further reading.[33]

War Department interest in the possible use of camels for riding purposes and as beasts of burden goes back at least to 1836. A study was made of the subject at that time by Maj. George H. Crosman and

No phase of Mojave Road history has captured the public fancy as much as Beale's famous camel experiment of the late 1850s. The scene above, called **"Camel Express,"** is a diorama believed to have been made in about 1938. Artist Carl Rakeman.
Courtesy, National Archives.

Edward Fitzgerald Beale (right) as he appeared in the late 1840s when he crossed the continent several times carrying dispatches for the government. From the book, *Memoirs of my Life,* by John Charles Fremont.

submitted to the proper authorities. Crosman's ideas were taken up as early as 1848 by Maj. Henry C. Wayne—an army officer whose name is closely connected with the subsequent history of the camel experiment. At about that same time the camels gained a legislative supporter in the person of Jefferson Davis, who was then in the United States Senate. Davis labored unsuccessfully until he left the Senate in 1851 to get the government interested enough in camels to make some sort of test.

In 1853, Jefferson Davis became Secretary of War in the Cabinet of President Franklin Pierce. He scored a victory for the camel experiment in March of 1855 when an appropriation of $30,000 was approved as an amendment to the military appropriation bill to "be expended under the direction of the War Department in the purchase and importation of camels and dromedaries to be employed for military purposes."

Davis secured from the Secretary of the Navy the loan of a storeship (the *Supply*) and one of the Navy's best officers—Lt. David Dixon Porter, Jr.—to import camels. In June of 1855 the ship sailed for the Middle East. The principal actors in the procurement phase were Major Wayne (who went along on this first voyage), Lieutenant Porter, and Gwynne H. Heap (Heap had formerly lived in Tunis and, as already pointed out, was an associate of Edward F. Beale). To make a long story short, the *Supply* arrived back in the United States on May 14, 1856, landing thirty-three camels at Indianola, Texas. The camels were marched to San Antonio and ultimately established at a new military post called Camp Verde, Texas, about sixty-five miles north of that town.

Meanwhile, Lieutenant Porter and Gwynne Harris Heap made a second trip to the Middle East and brought back another load of camels, landing at Indianola, Texas, on February 10, 1857, delivering forty-four additional animals to Major Wayne. Counting births and deaths that had occurred, this brought Major Wayne's total at Verde to something more than seventy camels.

During June, Beale assembled his party in Texas where they picked up the camels and the rest of their outfit. They left San Antonio for New Mexico on June 25, 1857, arriving in Albuquerque early in August. They left Albuquerque on August 12. Beale's instructions required him to commence his survey at Fort Defiance although, from a geographical point of view, Zuni was the logical starting point. To satisfy his instructions, and yet not delay the expedition, Beale himself went to Fort Defiance while he sent his train on direct to Zuni. Reunited with his train at Zuni late in August, Beale purchased corn from the Indians before setting out into the wilderness of northern Arizona. These preparations completed, the expedition finally got under way on August 31, 1857, heading west from Zuni.

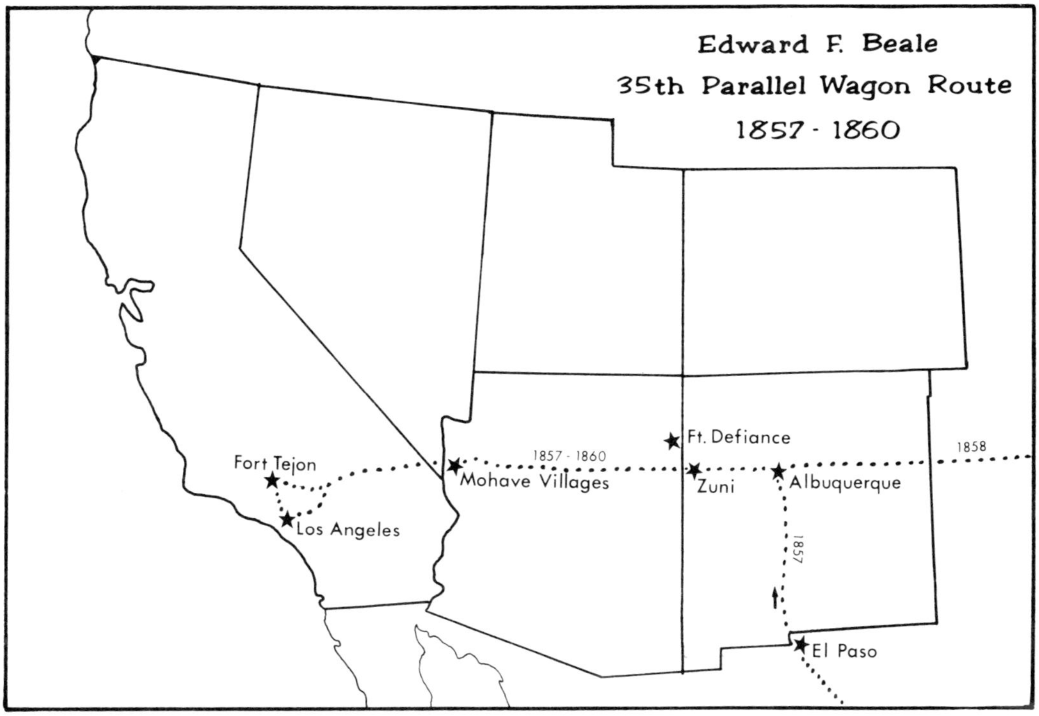

Beale had picked up a military escort at Fort Defiance. It was originally intended that he be provided with "one sergeant, three corporals, and twenty five picked men" for the trek through the dangerous Indian country of northern Arizona. Beale was given the option of taking a smaller number if he wished. This he did, taking with him an escort of only one sergeant, one corporal, and eighteen privates. These men were from Companies "B" and "H" of the 3rd Infantry stationed at Fort Defiance. Also, Beale borrowed five wagons with six-mule teams from the quartermaster at Albuquerque to carry supplies for the escort.

Leaving Zuni, Beale's expedition made a train of very respectable size, consisting of at least the following elements: eight heavy six-mule wagons, two ambulances, probably thirty-six or seven men in Beale's immediate party, twenty soldiers, twenty-five camels, over three hundred sheep, and a drove of more than fifty extra mules and horses for riding purposes and as spares for the teams.

Most of the military escort did not go through to California. On September 10 one man disappeared and apparently returned to Fort Defiance by himself and without taking leave of anyone. Members of the expedition thought he was lost, deserted, or crazy—but it appears he may not have been any of those things. The records at Fort Defiance show that one man had returned from the expedition before the end of

September and no disciplinary action was taken against him. Perhaps he had been lost. Then, at King's Creek, about one hundred miles east of the Colorado, Beale sent twelve of his escort with four of the army wagons back to New Mexico.

As had been the experience with earlier exploring parties, the trek through the eastern part of Arizona was easily accomplished. The expedition struck the Little Colorado River and followed that stream to the point where the Cañon Diablo enters and here they turned west toward the San Francisco Mountains. They were camped to the south of these mountains by September 12, less than two weeks after leaving Zuni.

The country to be feared was that to the west of the San Francisco Mountains. It was here that Beale succeeded where Whipple had failed. When faced with the desert country of northwestern Arizona, Beale faltered just as Whipple had done. But Beale took the necessary time to reconnoiter the country thoroughly and in the end was rewarded by finding his way through on the route that would later become the Atlantic and Pacific Railroad and U. S. Highway 66. This is roughly the stretch between Ashfork and Kingman, Arizona, by way of Peach Springs. To Beale goes the credit for finding the wagon route directly across northern Arizona that connected the pueblo of Zuni with the Rio Colorado of the West along the 35th parallel.

Leaving the vicinity of future Kingman, Arizona, Beale headed southwest across Sacramento Valley and crossed the Black Mountains through what is now known as Sitgreaves Pass. Beale called it John Howell's Pass and he seems to have understood that the Sitgreaves party crossed these mountains through what is now called Union Pass about ten miles farther north (Lt. Joseph C. Ives, Topographical Engineers, is the one who in 1858 misnamed the pass after Sitgreaves). Sitgreaves Pass is much rougher than Union Pass and there were places where Beale's wagons had to be let down by hand. One of his ambulances was smashed to pieces making the descent on the Colorado River side. The party arrived in Mohave Valley in the midst of the Mohave Villages on October 17, 1857.

Even with the loss of more than half his escort, Beale still had forty-three well-armed men when he arrived in Mohave Valley. The Mohaves would not risk an engagement against that force unless some fortuitous development gave them a sudden advantage.

Beale gave no indication in his journal entry that he felt threatened by the Mohaves. He wrote: "Here the Indians began to pour in upon us from the Mohave villages. First, two or three, and then by dozens. They were a fine-looking, comfortable, fat and merry set; naked excepting a very small piece of cotton cloth around the waist, and, though barefooted, ran over the sharp rock and pebbles as easily as if shod with iron." At another point he noted: "They are shrewder at a bargain, though, than our men, whose keen appetites cannot bear the delay necessary to a successful trade. The watermelons, cantaloupes, and pumpkins, are of excellent flavor and fair size." But the party did feel threatened, as reflected by Stacey's journal entry on October 19: "Expecting a fight all day–Men in readiness," and then the next morning: "Much astonished to wake up this morning and find my hair safe."

Beale, too, was concerned about the peaceful intentions of the Mohaves, whether he indicated this in his journal or not. As will be seen later, when he returned from California in January 1858 he brought along an entire company of dragoons to cover him while he passed through the Mohave Villages. And, on April 26, 1858, when he was back in Washington, D. C., he included the following paragraph in the letter that forwarded his reports to the Secretary of War:

> I regard the establishment of a military post on the Colorado river as an indispensable necessity for the emigrant over this road; for, although the Indians, living in the rich meadow lands, are agricultural, and consequently peaceable, they are very numerous, so much so that we counted 800 men around our camp on the second day after our arrival on the banks of the river. The temptation of scattered emigrant parties with their families, and the confusion of inexperienced teamsters, rafting so wide and rapid a river with their wagons and families, would offer too strong a temptation for the Indians to withstand.[34]

Subsequent events would show Beale's fears to have been prophetic.

Beale had now arrived at the western terminus of the wagon road he was instructed to explore. He had been authorized by the Secretary of War to proceed to Fort Tejon should he be in need of provisions upon his arrival at the Colorado. Consequently, he crossed his command and left the Mohave Villages for California on October 21, 1857. It is unfortunate for us that neither Beale nor Stacey continued their journal entries across the eastern Mojave Desert in California. Beale closed his account of his westward trip with the following statement: "Crossing my wagons over the river on the common air beds which I had brought for the purpose, and use of which I recommend to others, I followed the United States surveyor's trail from the river to Los Angeles, my wagons and train taking the right-hand road, and coming directly from the Mohave [River] to the Fort Tejon." The term "United States surveyor" is believed to refer to the surveyors of the General Land Office who used this trail in 1854-1856.

It is known that Beale traveled up the Mojave River to approximately where Helendale is now located, at which point the wagon train turned northwest toward Fort Tejon. Two camels were detached from the train and taken into Los Angeles where they attracted a great deal of attention. Beale was criticized in the California Press for taking his train to his ranch near Tejon instead of moving it on into Los Angeles. One disgruntled citizen in Los Angeles wrote: "Lieutenant Beale has used the national dromedaries to build a road to his own home, and he will be alone benefited by the location of it." Apparently then—as now—there were many who did not appreciate the fact that Beale's official wagon road work stopped at the Colorado. Also, it was an empty criticism anyway since the road from the Mojave River to Tejon had existed for many years as did the one from the Mojave River to Los Angeles. Both routes were well-established and Beale's selection of one or the other could not have affected the future flow of commerce to any important degree.

Beale was now in California and in a few weeks would start back over his newly marked trail. But, before continuing the discussion of Beale's work, two divergent but related topics must be introduced—these are the Mormon War and the 1857-1858 steamboat explorations of the Colorado River.

The story of the so-called Mormon War of 1857-1858 is mostly outside the scope of the present paper. However, that episode did result in the presence of Mormons in Mohave Indian country and the eastern Mojave Desert and these Mormons may have been partly responsible for the Mohave attack on emigrant trains in August of 1858. Therefore, it is considered useful to provide a few paragraphs explaining this episode in American history.

After suffering persecution in New York, Ohio, Missouri, and Illinois, the Mormons took refuge in the wilderness of Utah in 1847. They commenced the development of a desert kingdom. Their religious leader—Brigham Young—was also their political leader.

By 1857 there was a feeling on the part of many that the sovereignty of the federal government was not respected by the Mormons in Utah Territory. Brigham Young was the leader and he—and not the Congress of the United States—would decide what the laws were and how and when and against whom and in what manner they would be enforced. Further, gentiles who desired to settle or trade in Utah felt they were persecuted by the Mormons. There is no point in attempting to evaluate the claims made on both sides of these issues. Given the conditions under which the Mormons originally migrated to Utah, the climate that seems to have developed by 1857 is perhaps only what would naturally be expected.

The factor that finally pushed the government into action was the resignation of Federal Judge W. W. Drummond in Utah on March 30, 1857. Drummond's letter of resignation included a long list of charges against the Mormons and Brigham Young. This letter arrived in

Washington early in the new administration of President James Buchanan. Reacting to the information provided by Drummond and on the basis of counsel provided by others, Buchanan determined to install a new governor (non-Mormon) and other officials in Utah and to send along an army of twenty-five hundred troops to see that everything went smoothly.

The march against Utah was commenced in July of 1857 from Fort Leavenworth, Kansas Territory. During the fall and early winter, columns of troops and long supply trains moved slowly over the emigrant trails toward Utah. The Mormons reacted to these developments as though hostilities had been declared against them by a separate nation.[35] Brigham Young declared martial law in September and called out the territorial militia. The Mormons burned several forts on the roads that approach the Salt Lake Valley and they harrassed supply trains by using guerilla warfare tactics.

The army did not reach Salt Lake City before winter set in. They spent the winter of 1857-1858 bivouacked in temporary camps in the mountains. This turned out to be a blessing in disguise, as it gave both sides a chance to cool down. During this period, President Buchanan sent a secret envoy to the Mormons and differences were settled to the satisfaction of both sides through negotiations. The new territorial governor was accepted into office without the direct use of the army.

Meanwhile, in their desperate frame of mind, the self-reliant Mormons were considering all possibilities to guarantee survival. One course of action was that they would abandon their homes in Utah—burning everything to the ground—and relocate somewhere else as they had done so many times in the past. In pursuit of this policy, Mormons appeared in the eastern Mojave Desert and along the lower Colorado River late in 1857 and early in 1858. In addition to being on missions of exploration, they conducted reconnaissance activities to provide for early detection of any force that might be sent against Utah by way of the Colorado River route. Also, they made repeated attempts to turn the Mohave Indians against Americans.

Capt. Henry S. Burton, 3rd Artillery, who had been stationed for many years on the Pacific Coast, and who was as familiar with the Indians of the Colorado River as any man, wrote as follows about Mormon and Mohave relations during this period:

> About this time [early 1858], Mormon influence was discovered among the Mojaves, exciting them to commit depredations upon all whites who were not Mormons, visiting or passing through the country, and it was well known that small parties of Mojaves often visited the Mormon Missions, as well as that Mormons frequently came to the Mojave Villages during the Autumn of 1857. In the winter of 1857-1858 the principal Mojave Chiefs visited the Mormon Missions in the southern part of Utah.[36]

From other sources, three specific Mormon groups in Mohave country during this period can be identified.

The second side issue that must be introduced is steamboating on the Colorado River. Since early in the 1850s government supplies had been delivered at Fort Yuma by steamboats on the Colorado. Ocean-going sailing vessels would call at the mouth of the river and the small river steamers would off-load government stores and carry them up the river to Yuma. The steamboat business was owned and operated by George Alonzo Johnson.

For many years the possibility of navigating the Colorado above Yuma had been discussed. On a trip to Washington, D. C., Johnson had laid the groundwork for an appropriation to determine the navigability of the Colorado above Yuma.[37] In 1857 this project received increased interest because of the Mormon War. It was considered possible that an army, or at least supplies for an army, might be moved to Utah by way of the Colorado River route. If steamers could get up far enough an easy connection might be made with the Salt Lake Trail to Utah.

Somehow the appropriation went astray and the army Topographical Engineers received it instead of Johnson. Lt. Joseph Christmas Ives–the man who had been first assistant to Lieutenant Whipple in the Pacific Railroad survey over the 35th parallel–was assigned responsibility for conducting a survey to determine the character and navigability of the Colorado River above Yuma.

In June of 1857, Ives began construction of a special steamer. It was built in sections in Philadelphia and subsequently shipped to the mouth of the Colorado for assembly. The boat was fifty feet in length, made of iron, and fittingly christened the *Explorer*.

The *Explorer* was shipped to San Francisco via Panama. It was then shipped from San Francisco to the mouth of the Colorado and unloaded there early in December, 1857. At this time, Beale had completed his first westward trek and was sojourning at his ranch near Tejon. Also, the Mormon War was in full swing and Mormon scouts were among the Mohaves counseling them as to their future behavior with respect to the whites.

Meanwhile, steamboat man George Alonzo Johnson was understandably disgruntled over having missed a fat appropriation and the chance to be the first to explore the navigability of the Colorado above Yuma. He visited Ives at the mouth of the river while the *Explorer* was being assembled and had an interview with that officer that left Johnson feeling all the more cheated. As a result, he determined to take his own steamer–the *General Jesup*–and go on up the Colorado on a private trip of exploration, thereby depriving Ives of the pleasure of being the first to do this. Also, he managed to get the commanding officer at Yuma to detail an officer (Lt. James L. White) and fourteen men of Companies "A," "F," and "G," 3rd Artillery, to accompany him as escort. With this military addition, the objectives were added to check on the attitude of the Indians along the Upper Colorado and on Mormon activity in that area. Thus, Johnson's objectives included, perhaps on a reduced scale, all of what Ives was supposed to do.

The Johnson expedition left Yuma on December 31, 1857. A few weeks later they arrived at their high point on the river at the lower end of Cottonwood Valley nearly two hundred airline miles from Yuma (by the course they had to travel the distance was probably over three hundred miles). They were some twenty-five miles above Beale's Crossing.

Traveling back down the river, Johnson stopped near Beale's Crossing to take on wood. Much to his surprise, suddenly a large party of men appeared over the road from California. This was on January 23, 1858. The party was Edward F. Beale traveling from Fort Tejon and heading back east over his recently surveyed route. He was escorted by Company "F" 1st Dragoons commanded by Maj. George A. H. Blake. The other officers present with the command included Lt. John T. Mercer, Lt. Charles H. Ogle, and Dr. Peter G. S. TenBroeck, all attached to the 1st Dragoons.[38]

Beale's impressions of this chance meeting with Johnson as recorded in his journal have been repeated several times in published literature. In general, I avoid presenting that which is already available in published literature. In this instance an exception should be made and Beale be permitted once again to describe this dramatic meeting:

> *Saturday, January 23, 1858.* We reached the Colorado river early in the morning, having encamped in a rainstorm the night previous a few miles from it. Shortly after leaving camp, my clerk, F. E. Kerlin, who with two of my party had been despatched the day previous in order to have my boat ready for crossing, was seen returning. Various surmises were immediately started as to the cause, and as soon as he was within speaking distance he was questioned eagerly for the news. He gave us a joyful surprise by the information that the steamer *General Jesup,* Captain Johnson, was at the crossing waiting to convey us to the opposite side. It is difficult to conceive the varied emotions with which this news was received. Here, in a wild, almost unknown country, inhabited only by savages, the great river of the west, hitherto declared unnavigable, had, for the first time, borne upon its bosom that emblem of civilization, a steamer. The enterprise of a private citizen had been rewarded by success, for the future was to lend its aid in the settlement of our vast western territory. But alas! for the poor Indians living on its banks and rich meadow lands. The rapid current which washes its shores will hardly pass more rapidly away. The steam whistle of the *General Jesup* sounded the death knell of the river race.

Accompanying Captain Johnson, was Lieutenant White, of the United States army, and fifteen soldiers as an escort, which, with as many rugged mountain men, and the steamer as a fort, made a dangerous party to meddle with.

In a few minutes after our arrival the steamer came alongside the bank, and our party was transported at once, with all our baggage, to the other side. We then swam the mules over, and bidding Captain Johnson good-bye, he was soon steaming down the river towards Fort Yuma, three hundred and fifty miles below. I confess I felt jealous of his achievement, and it is to be hoped the government will substantially reward the enterprising spirit which prompted a citizen, at his own risk and at great hazard, to undertake so perilous and uncertain an expedition.

I had brought the camels with me, and as they stood on the bank, surrounded by hundreds of wild unclad savages, and mixed with these the dragoons of my escort and the steamer slowly revolving her wheels preparatory to a start, it was a curious and interesting picture.[39]

Lieutenant White had discovered that two Mormons had been among the Mohaves about twenty days prior to his arrival. White reported as follows about this activity:

In this valley [Cottonwood Valley] enters a trail from the Vegas, on the Salt Lake road. According to the account of the Indians, two Mormons had visited them by this route, about twenty days previous to our arrival.

I enquired very minutely in regard to the nature and object of this visit but could learn nothing very definite. The Indians stated that the Mormons who came among them, told them they were Mormon soldiers: that they wished to be friends with the Mohaves, and wished the latter to entrust to them the rearing and education of their children. At the last camp made by us, at the valley just mentioned, our Yuma guides learned from the Indians there, that the Mormons had armed the Pi-Uts Indians with guns and Pistols. That the Pi-Uts had told them that they and the Mormons were going to war with the Americans. These Indians then wanted to know of our guides whether or not the Pi-Uts' story were true. As our guide was uninformed of the state of affairs between the Mormons and the government, he referred them to Capt. Johnson and myself. Hitherto I had spoken not one word to the Indians about our impending difficulties with the Mormons: for the reason that I had not seen the Mohave chief. Finding however that the Indians at this remote point were better acquainted with the transactions of the Mormons and held intercourse with them, and with their Indian allies, and the opportunity being a good one, there being representatives from both the Cohualch and Mohave bands, there present, I determined to disclose to them, in a manner as comprehensible to them as possible, the full and, as I believed, truthful statement of the origin and cause of the impending hostilities with the Mormons. I advised them in case hostile engagements should occur between government troops and the Mormons, to take no part whatever in them; to espouse the aide of neither one nor the other side. But, if any Indians should join with the Mormons against our government, they would show themselves good friends to us by driving such away from their lands, should they go there for the purpose of annoying our people or troops. They were apprehensive that the Mormons would attempt to punish them for not yielding to their wishes as they had already threatened to do. But said if any Mormons should visit their valley again, they would immediately dispatch an express to Fort Yuma with the intelligence.[40]

The two Mormons referred to by the Mohaves as having been in their country about twenty days prior to the arrival of the Johnson party, were Dudley Leavitt and Ira Hatch. They had been sent to the river by Jacob Hamblin to investigate rumors that had arisen since the Mountain Meadows Massacre that Utah was to be invaded from the direction of California or the Colorado River. Leavitt and Hatch met with little success among the Mohaves on this trip.

Lieutenant White and Beale discussed the Mohave problem. The Mohaves had shown threatening signs to Beale's party the previous October–and yet they had committed no positive act of hostility. Therefore, the two men agreed to give the Mohaves the benefit of the doubt for the moment and consider them peaceful.

Leaving Beale's Crossing, Johnson steamed back down the Colorado. Nearing Yuma, the *General Jesup* struck a rock and sunk (it was later repaired and placed back in service). Johnson and his crew had to endure the embarrassment of arriving back at Yuma on foot. Still, to Johnson goes the credit to be the first to investigate the navigability of the Colorado above Yuma. It is a situation strikingly similar to Aubry's exploration of the 35th parallel in 1853 and 1854. In both ventures a private citizen risked his life and his own resources to carry out an act of exploration. In both ventures the explorers preceded the federal government in important spheres of exploration. And, in both ventures their achievements have been masked by elaborate government reports documenting in minute and colorful detail the exploits of later government engineers. Through the efforts of Arthur Woodward in his book, *Feud on the Colorado*, recognition of the steamboating achievements of Johnson and White has been gained. More recently, Don Chaput has focused his energies on a book-length biography of explorer Aubry, so that he too is taking his well-deserved place in history as a citizen explorer.

After being ferried over the Colorado by the *General Jesup*, Beale's party (less the dragoon escort and camels, both of which returned to California) continued their journey eastward over Beale's wagon road. A quick trip put him at the eastern terminus of his road in New Mexico on February 21, 1858. From there he continued on to Washington, D. C. By April 26, 1858, he was able to transmit his report to the Secretary of War. His submission consisted of a forwarding letter, his journal, and a fairly good map based on the one resulting from the Whipple survey. Beale's report was submitted to the House of Representatives in May and during the same month it was ordered to be printed—thereafter it was available to emigrants.

Someone in Major Blake's command—probably Dr. TenBroeck—provided a newspaper with impressions the troops had gained of the Mojave Road:

> The command under Major Blake, which was an escort to the Colorado for Mr. Beale, have returned to Fort Tejon, and were not a little disgusted with the road to the Colorado, on account of deep sand in some places, rocks in others, no wood and water in others, and grass nowhere. The troops beat Mr. Beale twenty hours to the Colorado.[41]

Meanwhile, the Ives party with the little metal steamer *Explorer* started up the river on January 11, 1858. Because of the low state of water in the river—and perhaps also because of the poor design of the steamer—Ives made painfully slow progress. It was not until February 10, 1858, that he arrived in Mohave Valley. He found the Mohaves to be friendly. He passed on through Mohave Indian country to a point about thirty miles above where Johnson had turned back. At the mouth of Black Canyon the *Explorer* struck a rock and needed repairs. While the repairs were being made Ives and a few others took a small skiff and rowed on up through Black Canyon to Las Vegas Wash—Ives thought he had reached the mouth of the Virgin River. This is as far upstream as he went. He sent a small land party out to make sure there was a practicable wagon road connection between the head of navigation and the Salt Lake Trail. These important things accomplished, he turned back downriver to the Mohave Villages.

There were two divisions to the Ives party. There was the group with the *Explorer*, personally directed by Ives, and there was a land party which consisted of a large pack train of supplies under Lt. John Tipton, 3rd Artillery, which worked its way up the river considerably behind Ives.

By the time rumors of these movements reached the Mormon settlements in Utah, it seemed to the Saints that Ives himself might represent some sort of advance guard or reconnoitering team and the Tipton train was an army marching against Utah. To check out these rumors, four Mormons—with Jacob Hamblin as leader—traveled down into Mohave Indian country. They found Ives' party; and one of the

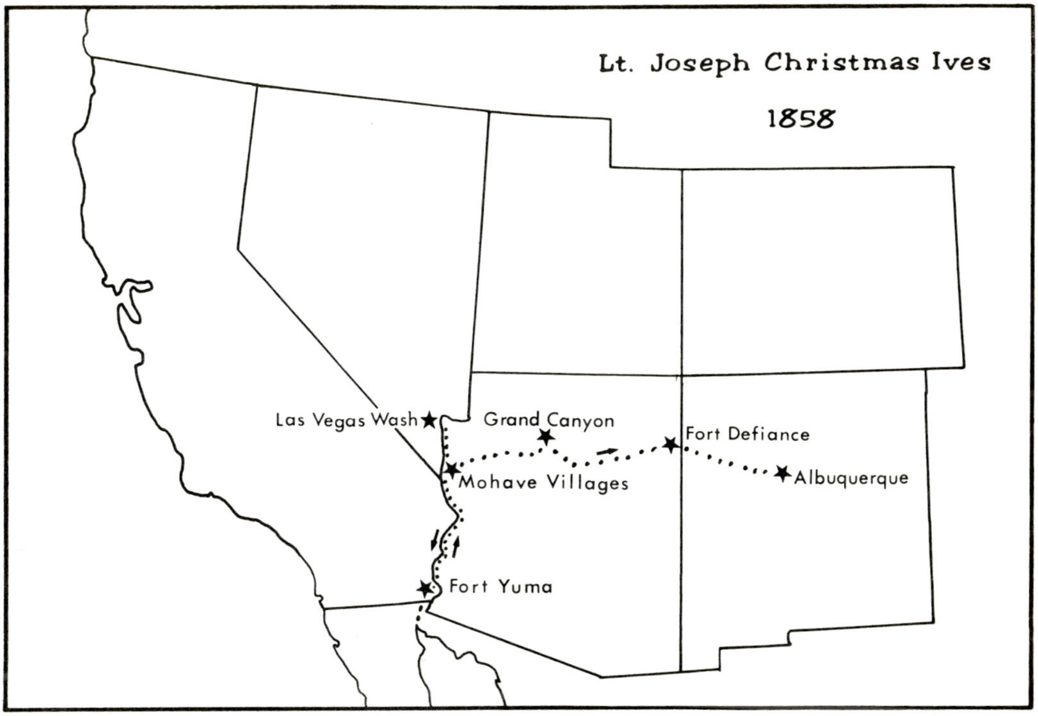

Mormons actually spent a night on the *Explorer*, having manufactured an excuse for being in the vicinity. Ives was well aware that his visitor was a spy.

These Mormons went among the Mohaves again with the result that when Ives traveled back down the river to the Mohave Villages he found the attitude of the Indians changed from friendliness to hostility. Ives stated as follows: "Those [Mohaves] who live in this neighborhood have suddenly become shy, and refuse to approach camp: Mariano has been among them, and reports that the Mormons have created a prejudice against us by informing the already suspicious Mohaves that we have come to take away their lands." I think this is an extremely important point. As will be seen later, the fear on the part of the Mohaves that the emigrants who came over Beale's wagon road later in 1858 intended to take away their land, might well have been the main cause for their attack on those trains. And it appears that this fear was planted in the Mohave mind in 1857 and 1858 by Mormons acting on one of the battlefields of the Mormon War. It may well be that the Mohave attack in August of 1858 was a result of the same policy of alliance-making with Indians that resulted in the Mountain Meadows Massacre the year before. Ives provides additional inputs regarding Mormon involvement with the Mohaves:

> The conduct of the Mojaves had been friendly till within a day or two, when their behavior became suspicious. They remained at a distance from camp, and at night attempted to set fire to the grass. Lieutenant Tipton had been strongly tempted to attack them, but felt reluctant to have any outbreak while ignorant of the condition of my party. Two Yumas, who had acted as guides, had a talk with the Mojaves, and told Mr. Tipton that the Mormons had been endeavoring in every way to excite the hostility of the last-mentioned Indians against the expedition, and had urged them to commence an attack by stampeding the animals. This statement coincides entirely with what Ireteba and Mariano have repeatedly told me. I have found these two Indians invariably truthful, and know not what object they could have had in manufacturing a false story. Corroborated as it is by the Yumas and by many circumstances that have occurred, I hardly know how to discredit it, though I feel reluctant to believe that any white men could be guilty of such unprovoked rascality. [42]

After considerable apprehension and negotiation, Ives managed to regain communications with the Mohaves and win their confidence. The point about the Indians' concern over their land is brought up again by Ives in the following statement wherein he is describing his reconcilliation with the Mohaves: "I further assured him [Mohave Chief Cairook] that he had been humbugged by the Mormons, and that so far from having designs upon the Mohave Valley, we intended to at once take our departure from it. Upon this he delivered to his people an earnest speech, which was respectfully listened to, and had the effect of altogether restoring harmony." On March 22, 1858, at the end of his sojourn in Mohave Valley, Ives wrote to the military authorities in California apprising them of his progress and the difficulties he had experienced with the Mohaves and Mormons. The part of his letter dealing with this subject is quoted immediately following:

> The Pai-ute Indians and a small party of Mormons have been for some time hovering about us. One of the latter came and passed a night in our camp, pretending that he and his companions were from the Utah settlements, but were not Mormons. He was very inquisitive about the Survey and its results. The party afterward went to the Mojave Villages and persuaded the Mojaves that we had come up to take their valley from them, but that the Mormons would be on the spot in a week, with the Pai-utes and destroy us. Our Mojave guides all deserted us and I thought it possible we might have some difficulty with their tribe, but I summoned their principal chiefs and endeavoured to remove their distrust. They now appear friendly tho' a little suspicious.[43]

With uneasy harmony restored, Ives divided his party on March 23, 1858, and proceeded out of Mohave country. About half of the party headed downstream with the *Explorer*, arriving at Fort Yuma on April 16, 1858. The other half–including Ives–took the pack train Lieutenant Tipton had brought up the river and started overland into Arizona to proceed on to Fort Defiance. One important event distinguished this trek across northern Arizona from those of Beale, Whipple, Aubry, and Sitgreaves. Early in April, 1858, Joseph Christmas Ives and his small band became the first white men to descend into the Great Canyon of the Colorado. Later, a contingent of the Ives group visited the Hopi Villages–which place had been left off the itineraries of the other explorers mentioned in this book–except Garces in 1776.

We must now direct our attention again to the Mohave Villages–the focal point for the history of the Mojave Road. A cycle in Mormon-Indian-American relations had been completed twice in just a few months. A Mormon party had been among the Indians in December creating a difficult situation for Johnson and White to deal with. Then another Mormon party came in March, creating a similar situation (although it was much more successful from the Mormon point of view than the earlier effort) confronting Ives. This process illustrates very clearly how readily the Indian attitude could be swayed by either Mormons or Americans. It seemed to be a situation where whoever talked to the Mohaves last had them as an ally–or, at least, had them alienated against the other party. So far as the steamboat explorers were concerned, the Mormons were to have the last word. A third party of Mormons came among the Mohaves after Ives left. So far as I can determine, this is the last group of white men to see the Mohaves before August when they attacked the emigrant trains.

By April of 1858 negotiations between the Mormons and Americans were proceeding positively in Salt Lake City and appearances were that a peaceful settlement would be found. Still, Brigham Young had a contingency plan to burn everything in Utah and flee to the south, rather than submit to the United States Army. Toward this end a party of nineteen men under Amasa Lyman headed south into Mohave country in April to select a good defensive site along the Colorado River. This party was among the Mohaves April 16-19. They found a site that seemed to serve their purposes and Lyman recorded in his journal as follows: "Noticed one place in particular, about 10 *ms* from our camp this morning, which affords excellent facilities for commanding the river, as the channel is, at this point, on this [west] side of the stream, and the adjacent rocks and ravines afford protection and shelter for a great number of men."[44]

Even though we have Lyman's journal, we do not know, of course, whether or not he counseled the Mohaves against the Americans. At first the Mohaves were cool towards the Mormons, but later Lyman was able to swing them back to their side. One of Lyman's final comments regarding the Mohaves seems to suggest that he had completely rewon their confidence: "Oat-sen-a-wantz [one of the chiefs] renewed his expressions of friendship and desires to have some of our people come and live with them, & assured me he would protect them."

On April 19, 1858, Lyman and his party left the Mohaves. They took the Mojave Road from the river to Pah-Ute Creek and from there went across the desert to inspect the Kingston Springs site, after which they returned to Utah and reported.[45] Back in Utah they discovered that a peaceful settlement had been made and the Mormon War was essentially over. No escape route to the south would be needed.

And now, after an unprecedented flurry of activity in their valley—featuring a variety of intrigues that must have confounded the mind of the untutored savage—the Mohaves would be alone for awhile. No more steamboats, no camels, no small bands of Mormons stealing down from the north, no wagons, no dragoons. Peace and quiet would return to the valley of the great Mohave Nation. But was the Mohave mind settled—were all doubts removed—or had seeds of suspicion and fear been implanted in his mind that would smolder and grow day-by-day? Would he watch his horizons anxiously for a sign that the numerous and powerful "Americats" did indeed plan to take away his lands as the seemingly friendly Mormons had repeatedly told him?

MOHAVES ATTACK EMIGRANTS

Early in 1858, events were taking shape that would put emigrants out on the 35th parallel during the summer of that year. Beale's expedition–and particularly the part involving camels–had been well publicized in the East. Parties about to leave for California would consider the possibility of going overland by this new route.

The most frequently traveled emigrant route to California went through Utah. For that reason, the prospect of troubles with Mormons in the Salt Lake country became a factor in selection of the 35th Parallel Route by California-bound emigrants. A later account of the trains of 1858, which was based on direct information from emigrants themselves, provided the following statement: "On account of the Mormon troubles in Utah, the emigrants decided to take the next route south of that territory, which would make their journey about 500 miles longer."

On February 8, 1858, in Decatur, Illinois (several months before Beale's report would be available), E. O. Smith wrote the letter quoted below to a congressman inquiring about the 35th Parallel Route. Smith's party would be one of those to try the route later that year. He was not in the fight with the Mohaves but he was a heavy financial loser in the venture.

> Mr. Gonin has just arrived at home from Washington he informed me that you would give me valuable information respecting the Waggon Road Rout near the 35' parlel Resently explored by Lutenant Beale. I desire to go that way. Expect to take mules Horses & Cattle about 50 of the former & 600 of the latter. Please send me such information as you can respecting the practability of the rout, anny thing respecting the road grass watter, Indians, Your experience no doubt will enable you to give me much good advice.[46]

Later, as the final results of Beale's expedition became known, articles appeared in newspapers extolling the virtues of the new route. The following article that appeared in the Santa Fe *Weekly Gazette*, May 1, 1858, was widely copied in the newspapers of the day:

> *The 35th Parallel Route.* We are informed by a letter from the Hon. M. A. Otero, our Delegate in Congress, that a large immigration is expected this year to pass over the route along the 35th Parallel, starting at Fort Smith, Ark., and passing by Albuquerque on to California. We published last week a very interesting and useful letter relative to the advantages of this route, from an experienced and reliable guide and traveller, who has passed over it frequently. It will be seen by that statement that it would be perfectly safe

> for parties small in number as six to undertake the trip. It is the safest, shortest and the best route in every respect for emigrants. We hope our exchanges in the States will republish this letter with a view to spreading the information it contains as widely as possible. It cannot fail to prove of incalculable benefit to those seeking the Pacific Country by overland travel.

There were at least five trains that attempted the 35th Parallel Route in the summer of 1858. Only two of these actually reached the homeland of the Mohaves. About their size, L. J. Rose (one of the leaders) reported as follows:

> The first company consisted of two parties: Joel Hedgpath, Thos. Hedgpath, G. Baily, Wright Baily, J. Holland Baily, John Udell, their families, and probably eighteen hands, forming one party, who had with them one hundred and twenty-five head of oxen and cows, twelve wagons, and fifteen horses.—Messrs. Bentner, Alpha Brown, S. M. Jones, myself and families, and seventeen hands, the other party, with two hundred and forty-seven head of cattle, and twenty-one horses and mules.[47]

In New Mexico the parties contracted with Savedra (a man who had accompanied both the Whipple and Beale expeditions) as a guide and left Albuquerque for California on June 30, 1858. Throughout July and August they struggled across northern Arizona. In many places they could not find Beale's trail at all. Some of the springs could not be located—others were dry or inadequate. The result was that when they reached Sitgreaves Pass and beheld for the first time the Colorado River of the West, provisions were low and their stock was worn out. But they were cheered by the sight of the river and the fact that the mountains on the opposite side, off in the distant desert, were in the Golden State. All they had to do now was cross the great river and proceed on a little more than two hundred miles to San Bernardino. And they had the prospect of trading with the "fat and merry" Mohaves, as Beale had described them, for agricultural products that the emigrants so badly needed.

Having crossed Sitgreaves Pass (or Beale's "John Howell's Pass") the emigrants were within a short march of the river. However, some members of the Udell/Hedgpath train did not feel their animals had the necessary strength remaining to pull their wagons to the river. Consequently, they unhitched and sent their animals ahead while they remained with their wagons and families at the pass. Most of the Rose train moved on to the river, arriving there on about August 27, 1858.

The emigrants did not find the Mohaves to be as merry as they may have been led to believe by Beale's report and by their own vain hopes to expect. Upon meeting the Mohaves, Rose reports: "They

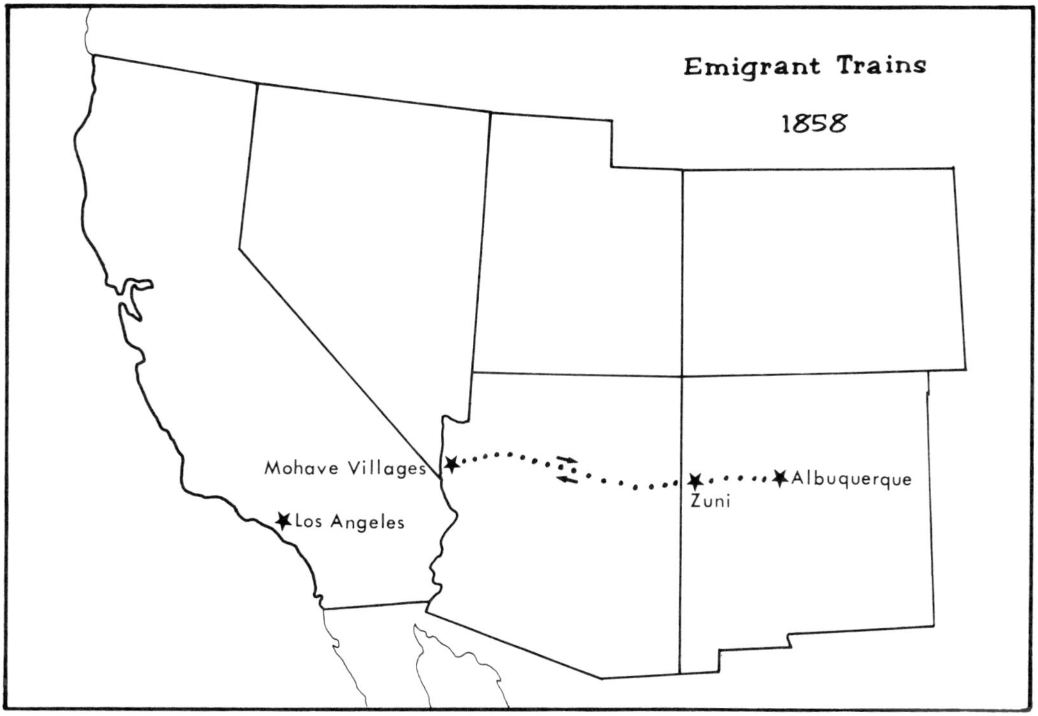

asked us how many of us there were, and whether we expected to settle on the Colorado. (Rather an unusual question I thought at the time.)." Later, when visited by a chief, the same question is indicated in the following quotation from Rose's account:

> About noon, one of the chiefs came to camp to pay us a visit. He had a number of his men with him. When told about his men stealing some of our cattle, he spoke to some of them, but evidently not very earnestly, and gave us but little satisfaction, only saying they would do so no more. He said but little, but was very attentive to all that was said or done and there was a cunning twinkle in his eye which I did not half like. He asked whether we were going to settle there, and when told that we were going to California, he gave us a very searching look, as if not half believing it.[48]

On August 30, 1858, the Mohaves launched a surprise attack against the emigrants at the river. Apparently the Indian plan was to attack the camp with a shower of arrows and then rush the group and finish them off with war clubs. The laxity that existed in the emigrant camp might have permitted this to succeed had it not been that the advancing Indians were noticed by a small girl playing in a wagon. She gave the alarm in time for the emigrants to obtain their arms. Although faced with three or four hundred Indians, bows and arrows were no match for muskets and Colt's pistols. The emigrants fought the Mohaves off. Twenty or twenty-five emigrants were badly wounded–but only one was killed.

Meanwhile, one wagon belonging to a German named Bentner, which had been near Sitgreaves Pass with the Udell/Hedgpath group, had started out alone to join the Rose train at the river. This wagon was attacked and all were massacred–Bentner, his wife, and five children. This brought the total number killed to at least eight.

Even though most of the emigrants had saved themselves from being massacred, the Indians had succeeded in driving off nearly all their stock. This, as has been related, represented the stock for both of the lead trains–not just the one that had moved its wagons to the river. That night the emigrants retreated back to the mountains where the Udell/Hedgpath wagons and families were waiting.

At about this time a third emigrant train appeared at the pass from the east. A general council was held and it was decided they would all retreat back to New Mexico. Had they known the Mohaves better, they would have realized they were more than strong enough to face them militarily and cross the river and proceed to California, which was less than half the distance to New Mexico. But this they did not know and so they turned around and headed back.

Traveling back across Arizona they met two more trains (one of which was the E. O. Smith group mentioned earlier) and each time provisions would be divided and the group of half-starved refugees was enlarged. Once the news of their plight reached New Mexico, the army sent out wagons of provisions, and other arrangements were made for their relief. During September and part of October they enacted one of the worst tales of hardship and suffering in the history of the frontier West.

Meanwhile, back at the Mohave Villages, the Indians probably felt they had successfully defended their homeland. The native mind could not have comprehended the power of the white man and the determination with which he would protect and avenge depredations conducted against his kind. The only thing Mohaves had accomplished was to hasten the day of their own destruction. Whatever the cause of their attack, they had committed a depredation of a type and magnitude that could not be ignored. By their action against these emigrant trains, they had initiated a process which in one year would result in the end of the Mohave Nation as an autonomous entity. Within a year hundreds of soldiers would come to their land, a fort would be established in their midst, and many Mohaves would be killed. And, most important to the focus of this paper, the ancient Mohave Indian Trail–the path of Garces, Smith, Wolfskill, Whipple, Beale, and others–would become a major wagon road.

THE COLORADO EXPEDITION

Emigrants were not the only ones to put Beale's Wagon Road to the test in 1858. During the late 1850s the Post Office Department experimented with various routes connecting the eastern states and the Pacific Coast. In 1858 a contract was made for a mail, frequently called the Central Overland Mail, over Beale's new wagon road. The terminal points were Kansas City, Missouri, and Stockton, California. From Kansas City to Santa Fe the route followed the well-known Santa Fe Trail. From Santa Fe west it was to use Beale's new wagon road all the way to Fort Tejon in California. From Fort Tejon there were well-established roads up through California's Central Valley to Stockton. The contractors were required to carry the mail once a month in each direction. Service under the contract commenced on October 1, 1858.

The attempt to carry mail by this route was not even marginally successful. Probably not more than two westbound and four eastbound mails were delivered throughout the nine months the contract was in effect. The main reasons for failure were attributable to the Indians and to the fact that Beale's Wagon Road was only a faint trace marked out through nearly a thousand miles of mountain and desert wilderness. The first westbound mail arrived at Stockton on November 24, 1858. and it brought news of the Mohave attack on the emigrant train. The proprietors of the mail route represented to the army that their mail contract was not operable because of the hostility of the Mohaves.

It appears that the Military Department of California (Brig. Gen. Newman S. Clarke commanding, with headquarters in San Francisco) may not have received news of the Mohave attack until the arrival of this first mail over the 35th parallel. That seems quite unlikely, however, since there was news of the massacre in Los Angeles papers as early as November 6. In any case, once informed of the attack it was not something General Clarke could ignore.

General Clarke lost no time in making arrangements for an expedition against the Mohaves. Further, the establishment of a permanent post that would protect the point where Beale's Wagon Road crossed the Colorado was ordered. He issued preliminary instructions on November 26, 1858. He ordered that four companies of the 6th Infantry commanded by Maj. William Hoffman of that regiment establish a new post somewhere "on the Mohave" in accordance with instructions which were to be supplied later.

Meanwhile, the Secretary of War personally learned of the Mohave attack and sent the following orders dated December 1, 1858, to General Clarke:

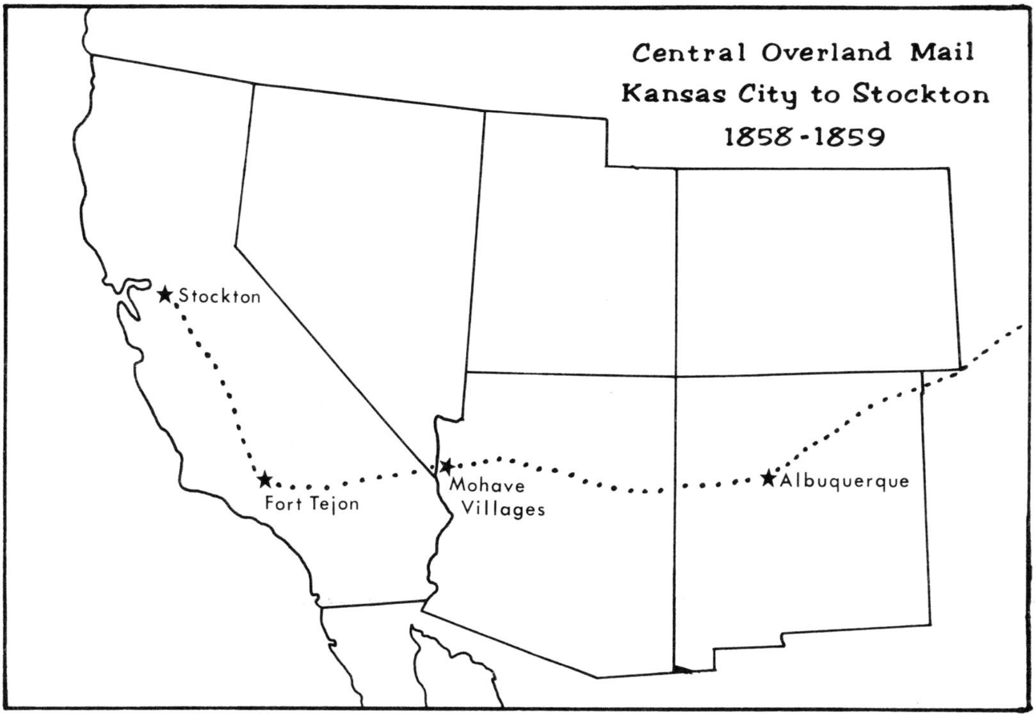
Central Overland Mail
Kansas City to Stockton
1858-1859
Stockton
Fort Tejon
Mohave
Villages
Albuquerque

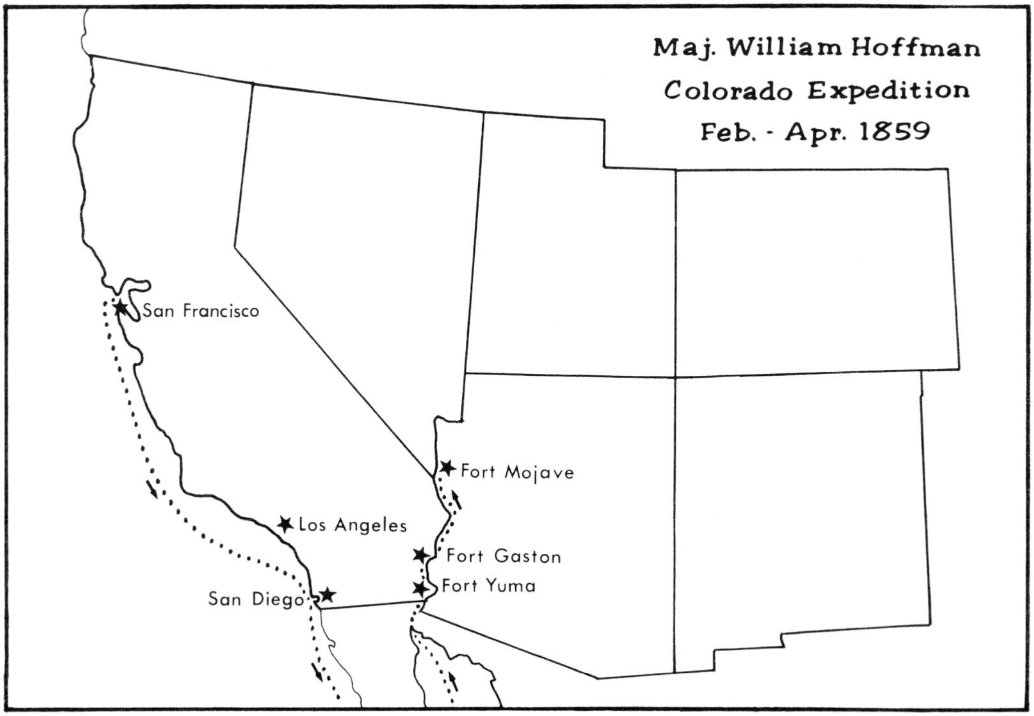
Maj. William Hoffman
Colorado Expedition
Feb. - Apr. 1859
San Francisco
Los Angeles
San Diego
Fort Mojave
Fort Gaston
Fort Yuma

The Secretary of War directs that you cause two foot companies to take post, with as little delay as practicable, at the point upon the right bank of the Colorado where the wagon road passed over by E. F. Beale, Esq in 1857, crosses that river.

The Command to be furnished with supplies for six months, and to be provided with two mountain howitzers and the necessary ammunition therefor.

The position to be occupied by these companies, and the route to be followed, are indicated upon the accompanying sketch.[49]

Those orders were slow in reaching California, of course, so in the meantime General Clarke proceeded with his own plans. On December 4, 1858, he issued specific instructions to Hoffman. The major was to proceed to the Colorado on a reconnaissance. He was to take one company of the 6th Infantry part way and with it establish a depot where the other companies might rendezvous later. At Martin's Ranch, near the mouth of Cajon Pass, he was to be met by a detachment of fifty dragoons from Fort Tejon, and that force was to serve as his escort for the reconnaissance to the Mohave Villages.

In accordance with these instructions, Hoffman took Company "E" of the 6th Infantry to Martin's Ranch and encamped them there December 27, 1858, to establish the advance depot. Then, accompanied by the escort of fifty men of Companies "B" and "K" of the 1st Dragoons, from Fort Tejon, he set out over the Mojave Road to conduct a reconnaissance in the vicinity of Beale's Crossing of the Colorado and to select a site for a post. He arrived at the Colorado on January 7, 1859.

Hoffman's plans to reconnoiter the vicinity of Beale's Crossing were frustrated by broken country, underbrush, and a multitude of Mohave warriors. From the accounts available, it would seem that these Mohave warriors may well have been more of a factor in the failure of Hoffman's reconnaissance than the broken country and underbrush. A clash occurred between Hoffman's command and the Mohaves and, although the Mohaves lost a few Indians killed and Hoffman had no casualties, it appears that no one was contesting the Mohaves' possession of the crossing when Hoffman left for the coast for a new set of orders and reinforcements.

Hoffman's attitude toward the Mojave Road was negative throughout his involvement with the expedition against the Mohaves. The following quotations, taken from three reports he issued following his January, 1859, reconnaissance to the Mohave Villages over the road, illustrate his position clearly:

The route over which I have passed to the Colorado I have no hesitation in saying is wholly impassable for trains, or for a larger body of troops than my escort; and any expedition against the Mohaves must be

> made by the river, if it be navigable, from Fort Yuma. A command should be sent of sufficient size, I would very respectfully suggest, to operate on both sides of the river at once; their supplies for a few days being carried on pack mules or camels, and a steamboat accompanying them to serve as a depot.[50]
>
> The route by land to the Colorado, over which I have just passed, is utterly impracticable for troops with trains in its present condition; and, even when all the improvements of which it is susceptible are made, in my opinion, the deep sands, which it traverses much of the way, will prevent the transportation of supplies upon it except at enormous expense; nor will it ever be used as an emigrant route, for the same reason. The troops, therefore, who are to establish the new post will have to reach their destination by Fort Yuma and the river; and, under this contingency, I very respectfully refer the matter to the general commanding the department for his further orders.[51]
>
> The instructions which met me at Camp Banning [near San Bernardino], prevented my completing the explorations which I intended making at these points, and I am unable now to say what difficulties would lie in the way of these plans, but this much is established beyond question, that all movements of troops against the Colorado Indians, with the land route in its present condition, must be by the river.[52]

It is difficult to understand Hoffman's position. He recommended that an extremely large command be sent against the Mohaves. More than eight companies of soldiers were ultimately sent under his command and even then he complained that he would not be able to handle the Indians because he could not have the support of two additional companies of dragoons from Tejon.

My assessment of this is that Hoffman—like the emigrants of August 1858—did not understand the Mohaves and their capabilities and limitations in warfare. The Mohaves could muster a large number of warriors—hundreds at least and possibly a thousand under favorable circumstances—but this force lacked firepower and could not be effectively utilized against an infantry company of the day. One company of infantry—as was later proven by Maj. Lewis A. Armistead in his confrontation with the Mohaves on August 7, 1859—could defeat the entire Mohave Nation. The river was the only factor that made it necessary to have more than just one company—this because there would be times when the company must be crossed over the river and this could mean that half would be on one side and half on the other. To cover that situation two companies were needed. In fact, War Department instructions pertaining to the establishment of a post at Beale's Crossing spoke of a two-company post.

The Mohaves themselves seemed to have had a clear understanding of their limitations. The record shows plainly that in spite of occasional very large numerical advantages, they would not risk an engagement with any large number of well-armed whites unless some special condition contrived to give them an unexpected advantage. But Hoffman did not understand all this and consequently called for ten companies when two would have been more than adequate. Hoffman's evaluation of the Mojave Road is another case of faulty analysis and poor judgment.

Before continuing a discussion of the complex events of 1859, it will be useful to devote a few words describing the status of the Mohave Indian Trail as it existed in January of 1859 when Hoffman passed over it.

When Garces first encountered it in 1776 it was a trail—or more probably a network of trails—that connected the Mohave Villages with other Indian settlements, with other trails, and most of all with the Pacific Ocean. Indications from the slight amount of data available indicate there was regular traffic on this prehistoric route. The impression is gained that on most any crossing of the eastern Mojave Desert the traveler would be likely to bump into one or two bands of three to five Mohaves who were off on a trading or visiting expedition. This use by the Mohaves seems to have continued right down to the 1850s. Although, with the rapidly increasing presence of white men on the California coast, Indian traffic over the trail had doubtless begun to diminish by the mid-1850s.

Since the time of Garces many white men (hundreds at least) had used this trail. An important feature of this early use is that generally the specific route was not used regularly by anyone. To almost all these early users it was a primitive trail through the desert wilderness. Everyone was just passing through—whether on errands of exploration or commerce or just travel—and it was not likely they would pass that way again.

This kind of use precluded the possibility that the route would become anything other than an Indian trail. To meet the requirements of an Indian trail, the springs need not be large and they could be far apart. To change that—that is, to locate additional springs or to improve those that were known—would require work. Little work of that kind had been done on the Mojave Road by early 1859 when Hoffman passed over it.

A few wagons had been taken over the route by Whipple, the surveyors of 1855-1857, and Beale—but not many. It would not be expected that an Indian trail would make a good wagon road without work. An Indian would go over a hill that a wagon would quite naturally go around. The expeditions involving the few wagons that had been brought over this route had accepted the route of the Indian trail rather than expend the days of work that might be required to find a better path for wagons. Hence, Whipple was willing to carry his little instrument wagon on occasion and Beale doubled teams, let wagons down steep hills with ropes, and carried part of his loads down a rough hill by hand.

So far as being a desert trail is concerned, the Mohave route presented no problem except that portion east of Soda Lake. West of that point the Mojave River provided abundant quantities of grass, wood, and water. Beyond Soda Lake was one hundred miles of "waterless" desert. Mohave guides had shown a number of white men how that desert could be traversed by passing from Soda Lake to Marl Springs, thence to Rock Spring, thence to Pah-Ute Spring, and on to the Colorado River. This was sufficient for small parties packing a few animals. But for a wagon road the following major problems demanded solution: a new route was needed around Old Dad Mountain to avoid an area of blow sand and the steep hill called Rocky Ridge, the water reservoirs at Marl and Rock Springs needed improvement, an alternate water source was needed near Rock Spring because it was frequently dry, an improved road was needed over Pah-Ute Hill. As will be seen, all of these essentials were accomplished during 1859 as the Mohave Indian Trail was transformed into a wagon road. Hoffman could easily have made these improvements with a tenth of the resources at his disposal—but he appeared never to have thought about a wagon road that *might* exist across the Mojave Desert. He apparently thought the Indian trail he traveled was the *best wagon road* the country had to offer.

Even if the Mojave route were as bad as Hoffman seemed to think, he still would have been immeasurably better off had he taken his command by that road than to ascend the Colorado River from Yuma. If the springs were a little too far apart on the Mojave route, he could *carry* water. Beale was using a water wagon in northern Arizona to support *his* explorations; surely Hoffman could find a water wagon in southern California. If the springs seemed to be a trifle small, Hoffman could attempt to improve them, locate others, and march his command in small detachments reforming at Pah-Ute Creek where there was enough water for the entire United States Army. If these possibilities occurred to him, he makes no mention of them.

It would seem that Hoffman's reaction to the Mojave route must have been caused largely by ignorance of desert country. Of course his unfamiliarity with the country is one of the reasons why he was permitted to hire two guides for the reconnaissance to the Colorado. He was accompanied by Joseph R. Walker (one of the most experienced and respected mountain men in the business) and William Goodyear (a man who had been involved in the recent land surveys of the eastern Mojave Desert). It would seem that these men would have explained the facts of desert life to Hoffman. Surely they did; but, if so, this is not reflected in Hoffman's reports or actions.

These were instances of poor judgment—and perhaps even lack of ability. But, behind it all, one gets the impression from many of the documents of this period that Hoffman did not want to command the Colorado Expedition. After his recommendations that the expedition should go by way of the Colorado River and consist of no less than ten companies, it was decided that General Clarke himself would exercise immediate command. But, after a few weeks, that was changed and Hoffman was returned to command.

Back in San Francisco, Hoffman, General Clarke, and Clarke's adjutant, Maj. William W. Mackall, made plans for the extensive expedition against the Mohaves. Hoffman's report left no alternatives as to the route to be taken—he had eliminated the possibility of using the desert road, the command must therefore be marched up the Colorado River from Fort Yuma.

It is clear from the communications emanating from department headquarters during the planning period that Clarke and Mackall doubted the soundness of the course they were pursuing. They knew very little about the route up the Colorado. One army officer under their control had marched men by that route. Lt. John Tipton had commanded the land party that marched up the Colorado a few weeks behind the *Explorer* in connection with the Ives survey.

Unfortunately, when the plans for the Colorado Expedition had to be firmed, Lieutenant Tipton was at faraway Fort Yuma and could not be consulted. Mackall wrote to him on February 1, 1859, as soon as it was decided the Colorado route would be used. His letter is quoted in full immediately following:

> You will have learned before this reaches you, enough to see that you are going in Command of some Howitzers with Hoffman up the River to 35th Parallel. Please write me all about the route. You know it. I would give a cool $100. to have you here now. Answer all my questions and give any other facts you know.
>
> Can mules subsist on the route without forage? If so how many?
>
> Tell me all about the road, character and length—distance from the River. make a sketch of your own—and mark the peculiarities as you go along.
>
> I don't want to go into a long letter but tell me all you know.
>
> In addition the state of the River above you. number of Indians &c.
>
> You know what questions you would ask. were you in my place. Answer them.
>
> Send your letter by Express. (Wells & Fargo's) to me, care of Banning Los Angeles and by the 1st Express.[53]

Mackall's letter has an element of desperation to it—and rightly so. Lieutenant Tipton (who on his march with Ives had found only four or five grazing places between Yuma and the Mohave Villages) could have set them straight on the virtues of that route. But Tipton's advice came too late. The wheels were set in motion to concentrate a large army and hundreds of animals and hundreds of tons of supplies at Fort Yuma.

At the time the decision was made to send this expanded expedition against the Mohaves, the units that would compose the Colorado Expedition were scattered all over California. In San Francisco were the headquarters of the 6th Infantry and Companies "C," "F," "H," and "I" of that regiment. Companies "B" and "K" of the 1st Dragoons were at Fort Tejon. Those two companies did not form part of the main body of the Colorado Expedition, but they served in a supporting role several times. Then, Companies "E" and "K" 6th Infantry were at Cajon Pass where Hoffman had left them after his unsuccessful reconnaissance. Company "G" 6th Infantry was stationed in San Diego. Finally, Fort Yuma was garrisoned at this time by Companies "F," "H," and "I" of the 3rd Artillery. To concentrate these units at Yuma posed a transportation problem of the first magnitude.

The first step in the movement of troops was to send Company "F" 3rd Artillery about fifty miles upriver from Yuma to establish a base for supplies. Company "F" arrived at the designated spot on February 14, 1859, and the new depot was named Fort Gaston.

Next, the two companies in Cajon Pass were marched across the desert over the southern emigrant route to Yuma. The trail they followed was the one being used by the Butterfield Overland Mail at the time. They left Cajon Pass on February 14 and arrived at Fort Yuma on March 2.

The units in San Francisco, together with many of the supplies for the expedition, were loaded on the steamship *Uncle Sam*. After one false start, the steamer left San Francisco for the mouth of the Colorado on February 16. A stop was made at San Diego where Company "G" 6th Infantry was picked up. This made five companies, plus Hoffman's headquarters, on the steamer. At San Diego about two hundred mules that were for the use of the expedition were put ashore. These were driven overland to Yuma. The steamer arrived at the mouth of the Colorado on February 27, 1859. By early March, therefore, all the elements of the Colorado Expedition were on the river although they were spread out over a distance of more than one hundred miles from Fort Gaston to the mouth of the river. It took Hoffman over three weeks just to get his troops and most of the supplies up to Yuma.

While Hoffman was struggling to concentrate his troops at Yuma, events were taking shape that would result in another clash between the Mohaves and white men before Hoffman could reach Beale's Crossing. Following his first wagon-road survey in 1857-1858, Beale returned to Washington and obtained another appropriation with which to continue and expand the work. In the fall of 1858, he had headed west from Fort Smith, Arkansas with a new assignment and appropriation. When cold weather came on (winter 1858-1859), he decided to rest his command in New Mexico until the worst of the winter weather had past.

During that period Beale made arrangements for his forthcoming march from Albuquerque to the Colorado River. His road-working party would consist of about fifty men and they would require quantities of provisions and other supplies. It would be advantageous to send some of these supplies from Los Angeles instead of carrying them westward with the road-working party. To implement this idea, he sent one of his men—Frederick E. Kerlin—from New Mexico to Los Angeles to procure the provisions and have them shipped eastward over the Mojave Road. Kerlin left Beale's camp in New Mexico on January 18, 1859 and, taking the overland mail stage from El Paso across southern New Mexico and Arizona, arrived in Los Angeles on February 2, 1859.

To insure the success of his assignment, Kerlin applied to General Clarke for an escort for the proposed supply train. At this moment the Colorado Expedition was being formed at Yuma; and, of course, those troops were not available for duty on the Mojave Road which was more than two hundred miles from Yuma. General Clarke's adjutant wrote to Kerlin and Beale on February 25, 1859, stating that an escort could not be provided because the available troops were at Yuma or were committed elsewhere. He strongly urged them not to attempt crossing the Colorado from either direction because of the danger presented by the Indians. He offered to allow Beale's party to draw supplies from Hoffman's command should they meet at the crossing of the Colorado.

It was decided by Beale's associates in California that they must make the attempt to carry provisions to Beale without a military escort. Samuel A. Bishop, Beale's partner in the cattle business on the Tejon Ranch, would command the expedition. They would travel over the Mojave Road and attempt a crossing of the Colorado at the usual spot in the midst of the Mohave Villages.

Bishop's party left Los Angeles on March 1. It consisted of about thirty-eight men, ten camels, six 6-mule wagons, and a number of pack mules. His ranks were swelled some by the addition of a mail party of the Central Overland Mail Company which joined him in Cave Canyon on the Mojave Desert. The mail party had been attempting to pass through Mohave country with the eastbound mail for several months.

On the way to the river Bishop's party took the time to scout a way around Rocky Ridge and the blow sand between Soda and Marl Springs. They dug a reservoir at the latter place thereby improving that water supply. Then the little band marched on to the Colorado River, arriving there on March 19.

For the next few days they attempted to cross the river. The force of forty-three men was strong enough to defend against the masses of Indians—estimated to have been as high as fifteen hundred on occasion—but Bishop could not risk dividing it into two parties as he had to do to effect a crossing. The Indians were more or less hostile throughout the time Bishop and his party were at the crossing, and a battle of some consequence resulted during which a number of Indians were killed. Bishop lost no men at the crossing, although one was wounded in the leg by an arrow.

On March 23, Bishop decided to return to Pah-Ute Creek where there were grass and water (and no Mohaves) and send two messengers to Major Hoffman asking for a detachment of men to assist him in crossing the river. While his messengers were gone (which was for nearly two weeks) Bishop put his men to work building a road over Pah-Ute Hill. More work would be required later, but for the moment they at least rendered it passable for wagons.

While these events were taking place, Hoffman was struggling to concentrate his companies at Yuma and making preparations to march up the river. General Clarke and Major Mackall had moved their headquarters to Los Angeles to be of assistance to Hoffman if he needed support of any kind.

In early March Clarke received a letter from the War Department informing him that he would be required to supply a large escort of dragoons to accompany a paymaster who was to take an army payroll to Utah. These dragoons would come from the two companies at Tejon. This put a wrinkle in Hoffman's plans—he had counted on having those two companies come galloping across the desert against the Mohaves from the west just at the time his other eight companies came steaming and marching up both sides of the Colorado from the south. Clarke wrote to Hoffman at Yuma informing him that the two companies would not now be available and that he should adjust his plans accordingly. Incredibly, Hoffman now felt the forces at his command might not be sufficient for the job at hand. He wrote: "I consider the diminished force now at my disposal, as inadequate to carry out my instructions in case of combined opposition on the part of the Indian tribes." Clarke replied immediately that he thought Hoffman's force *would* be sufficient.

With actions commenced that would result in subjugation of the Mohaves, General Clarke turned his attention to the question of whether or not a post was really needed at Beale's Crossing. After all, if the road were not good enough for the army to travel over to establish a fort to protect emigrants, then how would the emigrants to be protected travel over it? A good logical question–and Clarke acted on it. His orders for the establishment of a post at Beale's Crossing came from the Secretary of War and therefore permission must be gained from that office to change the plan. Consequently, while Hoffman was readying himself at Yuma and Bishop was fighting a thousand or more Mohaves, Clarke wrote the following letter dated March 21, 1859, to the Secretary of War recommending that no permanent post be established at Beale's Crossing:

> Since I undertook the establishment of a post on Beale's Crossing of the Colorado: facts have been brought to my knowledge which make me doubt the propriety of the movement at this time.
>
> How ever good the route on the 35th parallel may be to the Crossing of the Colorado, thence to the head waters of the Mohave it is a desert, and is 3 or 4 times that of the desert from Yuma.
>
> It is not likely that emigrants will follow the route, if once made aware of its true character; The post will be the most expensive post by far in this department–if I succeed in supplying it by water it will be at an expense of $300. per ton–if forced to do so by land the sum may be doubled.
>
> It was and is necessary to bring the Indians on the upper Colorado to submission, otherwise no white man would hereafter be safe in their Country and their impunity might influence those below, and render the lower route equally perilous–The expedition of Colonel Hoffman, was necessary whether he did or did not leave a garrison at the Crossing.
>
> This work I hope will be done in the time which will elapse before I can receive an answer to this, and that may be in time to prevent the building of a post if you should so decide.
>
> I can execute your instructions, but at so heavy an outlay that I feel it my duty to lay the facts before you.[54]

Finally, on March 26, Hoffman had his army readied at Yuma and commenced the march up the river. Bishop's messengers had hoped they would intercept Hoffman only a short distance below the Mohave Villages. Instead they had to go all the way to the depot at Fort Gaston–fifty miles above Yuma–before they intercepted the Colorado Expedition. Hoffman wrote back to Bishop that he did not want to divide his party (which then numbered nearly six hundred men) and that he expected to be in Mohave country in ten or fifteen days.

Bishop was in receipt of Hoffman's reply by April 5. With Hoffman's note in hand, Bishop had to make a decision. If he waited another ten or fifteen days, Hoffman might be at the crossing and he could proceed. But Bishop lacked confidence that Hoffman would arrive in that time. And, as it ultimately developed, Hoffman did require more than twenty days to reach the Mohave Villages and commence negotiations with the Indians.

In a letter of instructions to Bishop, Beale mentioned the route to the north of his wagon road that had been followed by explorer Francois-X. Aubry in 1853 and 1854. This route struck the Colorado far enough above the Mohave Villages that a crossing could be made without being opposed by a concentration of Mohaves. If Bishop could cross there he could push on into Arizona and meet Beale farther east. Wagons could not be taken because no real road was known to exist in that direction. Men on camels and mules could make the trip.

Bishop determined to take twenty of his men on camels and mules and carry some supplies to go meet Beale by this route. The bulk of the provisions he would cache at Pah-Ute Creek. A few men would be left there as a guard, and the remainder would be sent back to civilization with the wagons and teams. Bishop would go to meet Beale, and together they could work toward the Colorado. By the time they reached the crossing Hoffman would likely be there and they could recross and retrieve their supplies from the caches at Pah-Ute Creek.

Bishop's men worked hard on April 5 and 6, digging holes to cache the supplies. It was later stated that the labor of forty men for two days went into the work, implying the job was most thoroughly done. Then, on April 7, the wagons left for civilization; and on the same day Bishop and his men, mounted on camels and mules, departed for the unknown wilds of Arizona and the uncertainties of achieving a crossing above the Mohave Villages.

All went well for Bishop. On April 18, 1859, more than three hundred miles from Pah-Ute Creek, he met Beale near the San Francisco Mountains close to the present Flagstaff, Arizona. Bishop had effected a crossing of the river without incident and had emerged victorious from another skirmish with a band of Indians east of the river.

Hoffman was not idle either. His command had continued to inch its way up the Colorado and on April 20 was at the site of Beale's Crossing. Hoffman was finally among the Mohaves and ready for the great fight for which he had brought nearly six hundred men, hundreds of animals, and tons of provisions and equipment. He was to be disappointed in this, however, because—as might well have been expected–the Mohaves were completely overawed by the size of his command. They could not flee from their valley and they could not fight this formidable army. They took the only sensible option available to them and surrendered without further ado.

On April 23, 1859, a grand meeting was held between Hoffman and the Mohave chiefs, with the Colorado Expedition and hundreds of Indians looking on. Hoffman dictated his terms and, with very little quibbling, the Mohaves agreed to everything he asked. Hoffman's demands on the Mohaves were as follows:

> They must offer no oposition to the establishment of posts and roads, in and through their country, when and where the government chooses, and the property and lives of whites travelling through their country must be secure.
>
> As security for their future good conduct, they must place in my hands one hostage from each of the six bands.
>
> They must place in my hands the chief who commanded at the threatened attack on my camp in January last.
>
> They must place in my hands three of those who were engaged in the attack on the emigrant party at this spot last summer.[55]

A serious and important question can be raised about whether the Mohaves could have understood these terms because of the small amount of time devoted to the negotiations and because of the hopelessly complex translation procedure.

As to the time element, negotiations were essentially completed in less than one day. It seems unreasonable to assume that the ambassadors of two nations could work out the details of their future relations in less than one day. A realization of this objective would, of course, be impeded by the fact that the two parties had so recently been at war and that they had experienced only limited contact previously.

The language problem makes it seem even more unlikely that an understanding was reached. To carry on communication between Hoffman and the principal Mohaves, the following translation procedure was used: Hoffman made his statements in English–this was translated into Spanish by Capt. Henry S. Burton–this was translated into Yuma by a Diegueno Indian–this was translated into Mohave by a Yuma Indian. It can well be imagined that the words that came out at one end of the chain did not bear much resemblance to the original statements. The contention that the Mohaves did not understand the conditions imposed on them is borne out by subsequent events.

Hoffman was satisfied that he had solved the Mohave problem. He wrote optimistically to headquarters:

> I think it may be confidently relied on that they will never again as a body, give occasion for sending troops against them. Their whole demeanor from the time I entered the valley was that of a subdued people asking for mercy.[56]

FORT MOJAVE

And Another Battle With The Mohaves

Having completed these arrangements, Hoffman could make plans to return to the Pacific Coast with most of his command. He ordered Capt. Lewis Addison Armistead to remain at the crossing with Companies "F" and "I" of the 6th Infantry and a detachment of the 3rd Artillery to establish a permanent post. Hoffman called the post Camp Colorado but Armistead changed the name to Fort Mojave a few days after Hoffman left.

The bulk of the Colorado Expedition departed between April 26 and 28. Hoffman himself marched to the Pacific Coast over the Mojave Road to San Bernardino with Companies "C," "E," "H," and "K" of the 6th Infantry. Company "G" 6th Infantry and "F" of the 3rd Artillery, along with nine Mohave hostages, rode down the river to Yuma on the steamer *General Jesup.*

On May 1, Beale and Bishop with part of their road-making party arrived at Fort Mojave hoping for a fight with the Indians. Hoffman had deprived them of that pleasure through negotiations; and, to make matters worse, Hoffman also deprived them of their provisions—his men had stolen all of Bishop's caches at Pah-Ute Creek.[57]

The most significant happening, so far as the future of the Mojave Road was concerned, is that the army had established a post on the Colorado River at Beale's Crossing. Fort Mojave would require supplies. It had taken Hoffman much longer to march up the river than he had originally planned and consequently the garrison of the new post was very short on everything. One river steamer, the *General Jesup*, had accompanied Hoffman's command to the crossing and had brought some supplies. Another steamer, the *Colorado*, arrived at Mojave on May 12 bringing additional supplies.

The steamers—the *General Jesup* and the *Colorado*—had not navigated the Colorado with the dispatch that was expected of them. Partly this was caused by their design not being suitable for the shallow water and shifting sandbars of the upper Colorado. Another factor, perhaps, was that the contractor—none other than our old friend George Alonzo Johnson—managed to get Hoffman to approve an "open-ended" contract. When both steamers were engaged Johnson was paid five hundred dollars per day—three hundred for the larger steamer and two hundred for the smaller. The catch was that there was no limit set on the number of days it might take to go from place to place nor on the total number of days the contract could run. The result of the sandbars and poor contractual arrangements was that Fort Mojave could not be kept supplied by steamboat.

This situation gave rise to a second "open-ended" contract—this one for wagon transportation over the Mojave Road. Major Mackall, realizing that the companies of Hoffman's command that were being left on the Colorado would soon be out of provisions, contracted on May 1, 1859 with Phineas Banning to haul supplies to the Colorado over the Mojave Road. Banning was to use seven 10-mule teams, each carrying four thousand pounds of freight. He was to haul freight from the seaport of San Pedro to Beale's Crossing of the Colorado. His compensation would be $30.00 per day per wagon—$210.00 per day for the seven teams and wagons. As with the steamboat contract there was no limit on how many days it might take to get to Fort Mojave and back nor on the total amount that might be paid under the contract.

This contract resulted in raising more doubts about the practicability of the Mojave Road. Banning's teams left San Pedro on May 5, 1859, and did not return to that port until July 12. They were gone sixty-eight days! At $210.00 per day, that figured out to something like $14,280.00 to deliver a maximum of 28,000 pounds of freight to Fort Mojave. In round numbers that is fifty cents per pound—much more than was being paid by the river route even at the exorbitant rates being charged.

Banning's men did make some important improvements along the road. They did a little exploratory work discovering water sources and good grazing land on the Cima Dome, north of the road between Marl Springs and Cedar Canyon. Knowledge of this lush area as a place to recruit stock would benefit many future travelers on the Mojave Road. Second, they found Rock Spring to be dry and, as a result, they dug a well at a site about two miles west of Rock Spring that had a small seep and showed promise. They called their creation Banning's Well, but very soon the name was changed to Government Holes and the site retains that name today. This solved one of the basic problems to be overcome in changing the Mohave Indian Trail into a wagon road—that is, providing an alternate source in the event Rock Spring should go dry, which it occasionally did.

These were important improvements, but they were over-looked. The point that demanded attention was that it had cost fifty cents per pound to put freight down at Fort Mojave from the seaport of San Pedro. This seemed to corroborate everything Hoffman and other critics of the route had said about it. The fact that Banning had an "open-ended" contract and personally profited for every additional day consumed on the trip was not mentioned as a possible factor in the high cost. Attention was focused on the fifty cents per pound charged.

Meanwhile, Fort Mojave had become a permanent post. As mentioned earlier, General Clarke had written to the Secretary of War on March 21, 1859, recommending that a post not be established at

Beale's Crossing of the Colorado. His main reasons were that no emigrants were likely to use the road and also the expense involved in supplying the post. The following letter, dated April 30, 1859, at the War Department put an end to any hope Clarke had entertained of abandoning Fort Mojave:

> Your letter of the 21st ultimo in relation to the establishment of a military post at Beale's crossing of the Colorado, was duly submitted to the Secretary of War, who now directs me to say in reply, that after a full and careful consideration of all the facts and circumstances connected with the location of the post in question, he feels himself constrained to adhere to his original instructions on the subject, with the single modification suggested by you of substituting infantry for cavalry at that station.
>
> Should it be found that the transportation of supplies necessary for the new post cannot be obtained upon fair and reasonable terms, and much below the cost indicated, the Secretary of War will order the quartermaster's department to cause to be procured a small steamer of proper draft to convey the supplies from the mouth of the Colorado to the crossing.[58]

And the new permanent post would soon have work on its hands. Toward the end of June the next step was taken along the path to total and effective subjugation of the Mohaves when the hostages Hoffman had sent to Yuma escaped. It appears they had the impression they were to be imprisoned at Yuma only thirty days. The chief among them kept a string attached to his leg, and each day he tied a knot in the string. After there were thirty knots in the string, and it appeared they would not be released, the Indians became discouraged. On June 21, while they were out of their cells for their daily exercise, they broke loose. The old chief sacrificed himself by seizing the guard's gun. The others broke and ran. The chief was bayoneted and then shot through the head. The entire garrison was soon in pursuit of the fugitives. When the shooting had stopped, five of the nine Indians were dead. The remaining four escaped.[59]

This incident left both sides in an uncertain position. The idea behind keeping the hostages was that, if the tribe misbehaved, the hostages would be punished. But was this provision designed to work in reverse? If the hostages misbehaved, should Captain Armistead at Fort Mojave punish the tribe? The answer to this he did not know. He wrote to department headquarters for instructions. Meanwhile, he informed the tribe that they must give up the prisoners that had escaped. The Indians never complied with this demand.

The Indians were alarmed by the escape and by Armistead's demand. Clearly, they feared reprisals against the tribe. Suddenly there were fewer Indians around Fort Mojave. The all-important communications that Armistead had developed with Mohave leaders were broken off.

To this point very few trips had been made under the mail contract connecting Stockton, California and Kansas City, Missouri. The excuse cited by the contractors was hostility of the Mohave Indians. Now, with Fort Mojave established, the contractors were making one last effort to carry out the terms of their contract. They had a mail station on the river about two miles below Fort Mojave.

On July 20 and 21 the Mohaves stole stock from the mail station and finally attacked it outright. The employees of the mail company fled to the fort for protection. Then, a few days later, a routine guard patrol from the post was fired on. This was war again.

Armistead attempted to precipitate an engagement with the Mohaves, but they eluded his troops. On August 5 he successfully drew them into a battle and they were defeated. The battle was fought at a point about twelve miles below Fort Mojave on the east side of the river. Armistead had achieved a night march with twenty-five men, reaching the site of one of the largest rancherias on the river during the night of August 4-5. In the morning he surprised three Indians working in a field. One of these Indians was killed, thereby initiating the fight. Soon hundreds of Indians were involved. After a time Armistead's twenty-five men were reinforced by twenty-five more under Lt. E. G. Marshall. The arrival of this added detachment was in accordance with a previous plan. The Mohaves fought bravely, but were soundly defeated. Armistead counted twenty-three dead on the field. Other estimates indicate the number killed in the fight may have been much larger than twenty-three. Armistead had three men slightly wounded and none killed.

No overtures for a peace conference were made by either side. Armistead was determined to continue hostilities until the Mohaves were "satisfied." For several weeks after the battle, he sent out patrols hunting Indians and burning rancherias and destroying crops.

When Armistead's reports reached headquarters in San Francisco, the department commander became concerned for the safety of the garrison. Part of the dragoons at Fort Tejon were put under orders to march to his relief. Also, Company "E" 6th Infantry, then stationed at Camp Prentiss at the mouth of Cajon Pass, was alerted to prepare for the march. However, the Mohave War was over before that company could be sent to the scene of action.

By August 31, 1859, the Mohaves had enough. In the battle of August 5 they had tested their mettle against army troops in a pitched battle for the first time. The result was, for them, a complete disaster. Then, in the weeks that followed, they were given a sample of another type of warfare that white men could direct against their homes and crops. It was the end of the war and, perhaps unknown to the Mohaves,

Lewis Addison Armistead
Courtesy, Library of Congress

it was the end of their nation. Armistead's report of his meeting with the Mohave peace commission on August 31 was brief:

> I have the honor to inform you that I have, at their earnest solicitation, made peace, this day, with the Mohaves.
>
> I think they are sincere. They wanted a good whipping, which they got. They appear to be perfectly satisfied.
>
> All the terms that I could demand of them were that they should behave themselves, and give up the bend in the river, where the mules are herded and never come on it.
>
> I have done what I thought right, and if it only meets the approval of the commanding general, I shall be very much gratified.
>
> We shall be out of everything in the commissary department by the end of next month, September.[60]

It is more than a little ironic that Armistead forced the Mohaves to give up part of their land. It was the fear of losing land that had driven them to hostility to begin with. And now, in a very real way, their fears had been realized. Depriving the Mohaves of a part of their homeland helped crush their spirit, and this action marks the end of their existence as an autonomous nation. Armistead had been among them long enough to realize what the results of this condition would be. They could have taken any other form of punishment in their stride. Warfare, suffering, and starvation were specters the Mohaves had faced and survived before. But the white man's action in setting aside a part of Mohave Valley and requiring that no Mohave place his foot there without specific permission was a blow with which they were not prepared to cope. Their heritage and their physical power had failed. They were beaten.

It is sad when human beings are beaten so badly that they cannot continue the fight. But it is to the credit of the Mohaves that they had the wisdom to *know* they were beaten. To have continued the fight longer—as other tribes had done in similar circumstances—would have resulted in the death of many individuals (white men and Indians) with the same inevitable result. In 1859 the Mohaves had recognized the futility of their cause and started down the long and painful path of becoming part of the white man's world.

The map of what is now Arizona was changed in 1859. The earlier maps recorded a place called "Mohave Villages." After 1859 they had a place called "Fort Mojave."

WAGON ROAD AFTER ALL!

By the end of August, 1859, the Mohave Nation was subdued and Fort Mojave was firmly established in their midst. There was no emigrant traffic over Beale's Wagon Road and also the mail contract connecting Stockton with Kansas City had been terminated. There were essentially no civilians in Mohave Country except Beale's road-working parties and the military and their camp followers.

The big problems facing Fort Mojave were the building of suitable quarters and the obtaining of supplies. By the end of August supplies brought up by the steamers were nearly exhausted as were those brought overland by Banning's teamsters in June. Another steamer—the *Cocopah*—was on the way up and would arrive at the fort on September 1, but this would only delay the day of reckoning. The steamers could not keep the fort supplied and Hoffman and Banning had by their words and actions made the Mojave Road seem impracticable.

At this juncture a new individual stepped forward to play a decisive role in the history of the Mojave Road—Winfield Scott Hancock. Hancock was born in Pennsylvania in 1824. When only sixteen he received an appointment to West Point and was graduated from that institution into the regular army in 1844. He served through the Mexican War, had fifteen years in the army, and was only thirty-five years old in early May of 1859 when he arrived with his family in Los Angeles. As a young captain in the quartermaster's department he had been ordered to Los Angeles to receive all public property from Hoffman's columns as they returned from the Colorado Expedition. After that he was expected to provide for the flow of supplies from a quartermaster's depot he was to establish in Los Angeles to Forts Mojave, Tejon, and Yuma.

No one would have guessed in 1859 that in only four short years this young captain would wear a major-general's uniform and be placed in a position—on Cemetery Ridge at Gettysburg—where the very existence of the Union hung in the balance. He was catapulted to fame on that day, July 3, 1863, when he stopped Pickett's famous charge, thereby saving the day and perhaps the whole Union Cause. It was only a matter of time for the Confederacy after Hancock stopped Pickett's charge at Gettysburg. Later, in 1880, he would run for the office of President of the United States—losing the popular vote by less than ten thousand.

But, in 1859, he was a young captain with an assignment to provide supplies for faraway Fort Mojave. Hancock brought no prior knowledge of the Mojave Road with him. He had available the reports of Hoffman and other officers and the results of Banning's expedition (which was in process when Hancock arrived in Los Angeles). All these indicated that the land route to Mojave was impracticable for loaded wagons. Other records showed plainly that the fort could not be supplied adequately by the river steamers. Yet the Secretary of War had directed the fort was to be maintained at all costs.

Hancock hired several messengers to carry expresses from Los Angeles to Fort Mojave. These men were paid at the rate of about $100.00 per month plus a soldier's ration. They rode horses and carried the express (which included all mail intended for Mojave in addition to official mail) on mules across the desert on the Mojave Road to the fort. By this means Hancock kept himself informed as to the status of supplies at the fort.

Early in August, when supplies at the fort dwindled dangerously low because the *Cocopah* had not arrived as hoped, Hancock put together a train for Mojave consisting of two wagons, a small herd of beef cattle, and a pack train. The wagons were to be taken through nearly empty to insure that they could negotiate this impractical road. They were then to be left at Mojave for the use of the post. The expedition was considered to be an experiment to determine if Fort Mojave could be supplied by pack train.

On the Mojave River the train was joined by about forty dragoons officered by Lieutenants Benjamin F. Davis and LeRoy Napier from Fort Tejon. The large escort was sent along because at this time hostilities with the Mohaves were in progress and it was felt the train might need protection as it approached the Colorado and also Fort Mojave might need reinforcements. The train with escort arrived at Fort Mojave early in September–The Mohave War was over (for good this time) and the *Cocopah* had just delivered her stores at the fort. The two wagons were left at Mojave and the quartermaster's men returned to Los Angeles while the escort returned to Fort Tejon.

Back in Los Angeles Hancock questioned his packmaster (Alonzo Ridley, a respected Los Angeles citizen who, by the way, had accompanied Hoffman's Colorado Expedition as a wagon master) closely about the route. The determination and confidence with which Hancock subsequently supported the Mojave Road suggests he examined minutely the difficulties to be encountered.

There was heavy sand–how heavy and for how many miles? There were long stretches without water–how long? Could water be carried, new springs discovered? There was one bad hill–would it require rough-locking, doubling of teams, unloading the freight? The impression is gained that Hancock knew the answers to these questions and more before he started his famous train to Mojave on October 21, 1859.

By October the fort had consumed the supplies that had been brought by the *Cocopah* in September. They were in a dire strait again. It had been discussed that by a certain date it would be necessary to abandon the fort and march the troops toward San Bernardino while consuming their last remaining supplies.

On October 21, a train of ten 8-mule wagons left Los Angeles for Fort Mojave. They carried over twenty-five thousand pounds of freight for the fort and nearly fifteen thousand pounds of forage—most of which would be consumed by the teams on the trip. The wagons and teams belonged to the army quartermaster in Los Angeles. The teamsters were civilians hired by that quartermaster. The wagon master was Joseph Winston—a well-known citizen of Los Angeles who had accompanied Hoffman's Colorado Expedition. There were no "open-ended" contracts—the civilians were paid by the month and undoubtedly would prefer to draw their wages surrounded by the comforts of Los Angeles than in the desert wilds of the eastern Mojave. They would push their teams to relieve the sufferings of the garrison at Fort Mojave; and they would conduct a fair test of the Mohave Indian Trail (with recent modifications) as a wagon road.

This experiment was of the greatest commercial importance to Los Angeles. The government payroll in Los Angeles resulting from the quartermaster's depot ran about $2,000 per month—and that was big business in a small cow town in 1859. If the Mojave Road were found to be practicable, then supplies for Fort Mojave would be funneled indefinitely through Los Angeles; whereas, if they were to be carried by the water route, they would originate in San Francisco and be sailed directly to the mouth of the river. Los Angeles had a vested interest.

Winston reached Fort Mojave on November 5. He left Mojave for the return trip on November 7 and arrived in San Bernardino to pick up a load for Los Angeles ten days later. Winston reported: "The train employed ten teamsters and two herders, at $30 per month each. It was absent from this city [Los Angeles] thirty-three days. Two days were added by going to San Bernardino for loads for Los Angeles."[61] An article about Winston's trip and a narrative itinerary written by him were published in the Los Angeles *Southern Vineyard.* These are reproduced in the appendix to this book.

This experiment of Hancock's changed the adverse opinion of the Mojave Road that had been firmly established by Hoffman and others. As a result of this one expedition an Indian Trail was transformed, in the minds of men at least, into a wagon road.

This was only the first trip.[63] Hancock had demonstrated it was feasible to send loaded wagons by this route. He proved the soundness of his judgment by sending train after train to Mojave from Los

THE ROAD FROM LOS ANGELES TO FORT MOJAVE.

Mr. J. Winston, wagon master, returned this week from his journey to Fort Mojave, with Quartermaster's stores, and reports to Capt. Hancock, that he made the trip out in sixteen days, having a good road to travel over, and encountering no accidents or drawbacks. Mr. Winston returned from Fort Mojave to San Bernardino in eleven days. The road was measured by an odometer, and each day's travel carefully noted by Mr. W. as follows:—[62]

Oct.	21	To Mud Springs	28.42	28.42
	22	Cocomungo	12.45	40.87
	23	Martin's Ranch	13.08	53.95
	24	Old House in Cañon	7.45	61.40
	25	Foot of mountain	8.30	69.70
	26	First crossing Mojave	23.36	93.06
	27	Government Reserve	19.30	112.36
	28	Fish Ponds	20.85	133.21
	29	Lt. Davis' depot	23.00	155.21
	30	Cave	15.50	171.71
	31	Soda Springs	18.90	190.63
Nov.	1	First old crater	16.86	207.48
	2	Marl Springs	12.64	220.12
	3	Rock Springs	19.35	239.47
	4	Pi-Ute Creek	23.67	263.14
	5	Fort Mojave	22.57	285.71

Angeles during 1859, 1860, and 1861. After the first few trips he submitted a report to the chief quartermaster of the department wherein he estimated the cost of shipping freight by this route at four and one half cents per pound. He added optimistically that it might actually be obtained at less than that figure after the route became better known. He reported that this estimate included allowances for "hire of wagon masters, herders, and teamsters, their rations, the forage and shoeing of animals, and all repairs made each trip." Not included in the estimate was the "wear and tear of animals, wagons, and harness." A reviewing officer pointed out that 30% should be allowed for these items of expense that were excluded. But, even if the 30% was allowed, the cost still looked like only six cents per pound or even a little less. And even if that estimate was off another 100% it still appeared freight could be delivered at Mojave for around twelve cents per pound. This was a long way from the fifty cents per pound Banning had cost the government; and, more importantly, it was better than the steamboat companies were able to do for many years to come. As subsequent history would show, ten cents per pound was a typical charge to carry freight from San Pedro or Los Angeles to Fort Mojave.

The most important result of Hancock's experiment, so far as this paper is concerned, was to demonstrate without question that there was a practicable wagon route across the eastern Mojave Desert to the Colorado River. The Mojave Road became a line on maps. That line was labeled "wagon road." The Mojave Road would serve in that role for the next twenty years as one of the main life lines to developments that were commenced in the wilderness of the eastern Mojave Desert in California and northwestern Arizona.

Hancock's trains–frequently under wagon master Joseph Winston–passed regularly through the eastern Mojave during 1859, 1860, and the first part of 1861. Each trip added to the store of knowledge of the desert. Contact was had with the desert Indians–they were classified as weak, poor, persistent beggars; and, above all, treacherous. This description stuck with them for the next ten years. Trains and travelers through the eastern Mojave Desert had to be constantly watchful of Indians, even though they rarely ever saw one. So long as the party was not small and consisted of, say, ten well-armed men, and they stuck together, there would be no problem. Prudent travelers, therefore, rarely saw Indians. But let a weak team fall behind a few miles and it would soon be discovered that the Indians that had not been seen in months had been watching all the time. Such were the ways of the desert Indians.

The Mojave Road was shown to be practicable for wagons–still it was a rough road. There were some fairly steep grades, blow sand, and rocky ground between Los Angeles and Fort Mojave. Occasionally wagons would be upset or the load would come loose and part of it fall off. Upon arrival at Fort Mojave these instances would result in the convening of a Board of Survey composed of officers of the post. The job of the Board would be to determine the cause of any shortages or damaged freight. No blame was ever placed on anyone for these losses, but still the regulations required that formalities be attended to. This is fortunate for us, because many reports of these Boards of Survey still exist.

Typical of the problems revealed in these reports (mostly caused by upset wagons) are the following: three office chairs broken, boxes of candles broken open and the candles damaged, five bottles of pickles broken and worthless, one bottle of lime juice broken in transportation, and on one trip more than two-hundred pounds of sugar sifted out of the barrels in which it was being carried. To these losses could be added more than one thousand pounds of dried apples that were received mouldy and full of worms, more than two hundred pounds of ham "full of skippers, mice eaten, partially decomposed," and dried meat that had lost nearly half its weight because of the dry climate. With the aid of these examples, we begin to gain some appreciation of problems encountered in freighting supplies to Fort Mojave in the early years.

One item in these survey reports raises the prospect that some happy times were had along the hot and dusty Mojave Road. The manifest of a train that arrived at Fort Mojave on February 13, 1860, included two forty gallon barrels of whiskey. Presumably, since this appears in the official records, the whiskey was intended for the medical department. In examining the load it was discovered that one forty gallon barrel contained twenty-nine gallons and the other only about seventeen. It is easy to imagine that these spirits were not wasted away by leaks that opened in the barrels as the wagons bounced over the rough desert trail. The wagon master claimed the barrels were turned over to him deficient in quantity. He was supported in this by the Board of Survey.

The "**S. S. Senator**," one of the coastal steamers that connected San Francisco with San Pedro and other ports in southern California during the days of the Mojave Road. From an oil painting signed and dated "Joseph Lee, 1875." *Courtesy, California Historical Society.*

Los Angeles, as it appeared to the railroad surveyors in 1853 and 1854. From: *Reports of Explorations and Surveys, to Ascertain the Most Practicable and Economical Route for a Railroad from the Mississippi River to the Pacific Ocean* (Vol. V), by Lt. Robert S. Williamson and others.

Cajon Pass–One of the great gateways through the mountains connecting the East with southern California. In this picture Interstate Highway 15 sweeps majestically through the pass to the desert beyond. Etched in snow, high peaks of the San Gabriel Mountains form a backdrop.

Point of Rocks on the Mojave Road. A drawing from the 1860s by W. H. Hilton
Courtesy, Henry E. Huntington Library and Art Gallery

Remains of the Commissary and Quartermaster storehouse (above) at new **Camp Cady** as it appeared in about 1910. The adobe structure was destroyed in the 1938 flood. *Courtesy, O. A. Russell and Ken and Mabel Wilhelm.*

"The Caves" in Cave Canyon (now Afton Canyon) were a favorite camping place on the old Mojave Trail. The Mojave River is on the surface here and forage was usually available. The site was about sixteen miles from Camp Cady which was about a day's travel with loaded teams. For westbound travelers this was the first point where the reliable waters of the Mojave River were encountered. This picture was taken in 1967.

Dennis Casebier (above) standing next to one of the old rock ruins at the **Soda Springs** site.

In 1975 Soda Springs, or **Zzyzx Mineral Springs** (right), was deserted except for caretakers engaged by the Bureau of Land Management. Still, "Old Glory " was being raised over the historic site each day. Rippling in the prevailing southerly winds, the flag provided the only animation at this once busy site.

This picture shows a faint trace of the old **Mojave Road**. The view is taken to the west looking down into Soda Lake. The Soda Springs site is on the opposite side of the lake at the base of the mountains at a point just above the heads of the author's Chief Assistants. Photo taken in 1975.

Between Soda and Marl Springs the old road skirted along the west and south sides of this beautiful **cinder cone and lava flow** area. Picture taken in 1975.

A view (above) taken from the hill west of **Marl Springs** looking east. Part of the old relay station at the upper spring is shown in the foreground. The old Mojave Road runs from the springs diagonally off through the picture. Photo taken in 1966.

The bottom picture on the opposite page was taken at **Marl Springs** by R. D'Heureuse in about 1863. The comparison picture was taken in the same spot in 1975. The large stones showing in the older picture have been pushed to one side and hence are out of view in the more recent shot. *D'Heureuse photo, Courtesy, The Bancroft Library, University of California, Berkeley.*

Government Holes (above). A well was first dug at this spot in 1859 by teamsters in the employ of freight contractor Phineas Banning. The teamsters called their creation "Banning's Well." By 1860 the location was being called Government Wells or Government Holes and the site has retained that name to the present time. This picture was taken in 1967.

The view on the left was taken near Cedar Canyon looking west. The old road can be seen meandering through the Joshua tree forest on its way between Marl and Rock Springs. This picture was taken in 1967.

This soldier inscription is located near **Rock Spring.** For many years its meaning tantalized but eluded historians. Detailed research in military records in the National Archives has revealed that one Musician Charles Stuart of Company "B" 4th Infantry, California Volunteers, was at this spot on May 16, 1863. This was one of the companies sent to regarrison Fort Mojave during the Civil War. The fort was officially reestablished on May 19, 1863. Stuart's company—along with Company "I" of the same regiment—had spent the night of May 16/17 camped at Rock Spring on their way from Drum Barracks to Fort Mojave.

In long stretches in **Lanfair Valley** the trace of the old road has become deeply eroded and serves as part of the local drainage system. Photo taken in 1967.

Marlou and Darelyn Casebier (left) standing beside **Pah-Ute Creek.** This stream is a great enigma of nature. In one of the most barren and rocky corners of the great Mojave Desert, an estimated 250,000 gallons of water per day rise magically from the sand and begin a mile-long run through the Pah-Ute Mountains. Evidences of use by man from early prehistoric days until the present are visible along the creek. This picture was taken in 1973.

The army maintained an **outpost at Pah-Ute Creek** in 1867 and 1868 to protect the Arizona Overland Mail from roving bands of desert Indians. This view (below) was taken inside the ruins of the main fort building looking southwest toward Pah-Ute Pass. The notorious Pah-Ute Hill—one of the worst hills on the wagon roads of the West—is just to the left of the small dark peaks on the horizon. This picture was taken in 1973.

The three photographs on these two pages form a panoramic view of **Fort Mojave** that dates from 1871. The photographer was standing with the Colorado River to his back. The first picture (opposite page, top) is taken to the northeast showing the men's quarters. The second picture (opposite bottom) is looking east showing the flagpost, guardhouse, and storehouses. The third picture (above) looks to the southeast and shows the officers' quarters. The pictures were taken by Timothy H. O'Sullivan, photographer with the Wheeler survey in 1871. *Courtesy, National Archives.*

CAMP CADY
Trouble With The Pah-Utes

The powerful Mohaves were completely subdued with Fort Mojave established in their midst. They were agricultural Indians and not hunters. Therefore, they could not flee to the mountains and carry on limited guerilla warfare tactics as did so many Indians in their initial conflicts with white men. But the Pah-Utes living in the midst of the eastern Mojave Desert were different. They practiced agriculture to a small extent–planting small patches of corn, wheat, and squash at some of the larger springs. But largely they were nomadic, living off the land. They drifted with the seasons gathering nature's offerings of wild seeds, piñon nuts, lizards, tortoises, and other small game.

They appeared poor and wretched to early travelers in their country. When they approached passing trains it was as persistent and bothersome beggars. They impressed many early travelers as being arrogant. They would steal things left lying about a camp. Stock would disappear at times, no doubt providing a few days of feasting for these half-starved children of the desert.

From early times the desert Indians had shown that they would commit more serious depredations if the advantage should be markedly in their favor. For example, when Whipple left a herder behind to guard two exhausted mules, the watchful Indians murdered him and took the mules for food. Almost everyone that traveled through this country in the early days knew about the disposition of the Indians and consequently opportunities to commit these serious depredations did not come often.

While conducting the Colorado Expedition, the military was made aware of the hostile character of the desert Pah-Utes. At several points in the military correspondence of 1859 the commander of the Colorado Expedition, and later the commander of Fort Mojave, was reminded by higher authority to insure the peaceful character of the Indians in the desert. Once, in 1859, a column of Infantry from the fort marched westward to Pah-Ute Creek in pursuit of Indians but without success.

Generally, hostilities with the desert Pah-Utes originated from small bands of about twenty Indians. Only in a few instances are there records of these Indians assembling in greater numbers, thereby permitting them to attack large parties. The realities of the harsh desert environment in which they lived forced this result–a large army could not be supported for an extended campaign. Since their depredations were few and usually small in scope, efforts to control them were generally only half-hearted. An exception to this occurred early in 1860 when the Indians were credited with three murders.

The first incident was the murder of Robert Wilburn near the

Mojave River on January 23, 1860. Wilburn, an employee of Bachman and Company of Los Angeles, was guarding a herd of cattle on the river bottom. The Indians had stolen some stock. While searching for the stolen stock, Wilburn was ambushed and killed by some Indians who were assumed to be Pah-Utes. This created a feeling of alarm in Los Angeles with demands for revenge from the military. No action was taken immediately.

The second incident, and the one that led directly to the campaign against the Pah-Utes, was the murder of Thomas S. Williams and the wounding of his brother-in-law Jehu Jackman at Bitter Springs on March 18, 1860. In this instance a train of fourteen wagons belonging largely to Williams was wending its way over the Salt Lake Trail from Utah to Los Angeles to pick up freight for Utah. There were a number of teamsters and some passengers with the train. In the vicinity of Bitter Springs the train was joined by several Indians supposed to be local Pah-Utes. The Indians volunteered to show Williams a place where water and grass could be obtained. Williams and Jackman trustingly went ahead with them. The Indians attacked; Williams died that day and was buried at Bitter Springs; Jackman died a month later in San Bernardino.

This produced such a strong reaction in Los Angeles that the *Star* was moved to issue an "extra" on March 26, 1860. Petitions for a military campaign were circulated and signed in Los Angeles. One was sent to Governor John G. Downey (a resident of southern California) and one to General Clarke. As with the situation produced when the Mohaves attacked the emigrants in 1858, General Clarke had little choice but to take some action against the desert Indians.[64]

Clarke sent Capt. James H. Carleton with more than eighty soldiers from Companies "B" and "K" 1st Dragoons, to the Mojave country to punish the Pah-Utes. Carleton's instructions ordered him to punish any Indians he might find in the vicinity of the murders at Bitter Springs. Ultimately, he was given latitude to roam far and wide over the desert and to attack any Indians he might find.

The expedition left Fort Tejon bound for the Mojave River on April 12, 1860. In addition to Carleton, the officers of the command included: 1st Lt. Milton T. Carr, 1st Lt. Benjamin F. Davis, and Surgeon Jonathan Letterman. On April 19, 1860, the command reached a site on the Mojave River where the water comes to the surface about ten miles downriver from the point where the Salt Lake Trail branches off into the desert. Carleton established a base camp here. His men built a strong redoubt at the site, while he sent scouting parties into the desert to hunt for Indians. He named the new post Camp Cady in honor of his friend Maj. Albemarle Cady of the 6th Infantry who was at that time in command of Fort Yuma, California.

Carleton's troops remained on the desert until July 3, 1860, when they decamped and returned to Fort Tejon. In the time they were there they had ranged great distances over the desert. Detachments had been sent to Las Vegas over the Salt Lake Trail, into Death Valley, and east over the Mojave Road as far as Marl Springs. They had scouted south of the Mojave Road as far as Granite Mountains and, in fact, had passed completely around that range. They had examined much of what is now known as the Devil's Playground.

They had several skirmishes with the Indians as a result of which a number of Indians were killed. Carleton had the bodies of two Indians hung on a gallows at Bitter Springs and the heads of several others mounted there. He received a rebuke from department headquarters for this unpleasant display of human remains.

On July 1 a number of Pah-Utes came into Cady for a peace conference. Carleton explained to them why they were being punished and how they should behave in the future. They agreed to his terms and promised to be helpful to travelers and to commit no more depredations. He told them other soldiers would be out with presents for them in the fall if they kept their promises. If they did not keep their promises, he assured them that other soldiers would come and hunt them down just as his dragoons had done and that they would be severely punished.

Carleton's troops had erected three redoubts on the desert trails. The one at Camp Cady has already been mentioned. Another, called Hancock's Redoubt (named for the quartermaster in Los Angeles), was erected at Soda Springs about thirty-six miles east of Cady. The third was erected at Bitter Springs on the Salt Lake Trail.

As will be seen in this narrative, Camp Cady later became an important army post and it retained the name Carleton had given it during his brief sojourn in 1860. Soda Springs played host to a small outpost again in 1867-1868, but the name Hancock's Redoubt was by then forgotten. Bitter Springs had no future as an army post beyond the short use made of it by Carleton's troops.

Carleton's Pah-Ute Campaign brought more men, freight wagons, and traffic of all kinds into the Mojave Desert. And, there was still the steady flow of traffic between Los Angeles and Fort Mojave.

In the fall of 1860 the Indians gathered on the Mojave River in anticipation of Carleton's promised return. They stopped the express rider on his run between Los Angeles and Fort Mojave and told him they were waiting to see Carleton and that he had promised to be there at about that time. They threatened to attack Joseph Winston's train the next time it passed through from Fort Mojave. They said they thought Carleton was afraid to come among them again.

This news was conveyed to Carleton. His remark to his superior officer was: "I am not a volunteer for the position of commander of any body of troops that may be sent to council with, or to fight the Pah-Utes." No troops were sent to the Mojave River in the fall of 1860. As the year drew to a close, men in the United States ceased to worry about the activities of the lowly desert Pah-Utes. Instead, such issues as secession, slavery, and the prospects of Civil War occupied their attention.

In January of 1861 a situation occurred which portrayed—in a way—an almost complete reversal of roles for the white man and Indian over the Mojave Road. For at least eighty-five years the Mojave Road had been the domain of the Indian. Repeatedly he had guided white men over the route. Garces, Smith, Whipple, and others had used Mohave Indian guides to cross the desert on the ancient trail.

In the month indicated, however, the tables were turned. The Mohaves were curious about the land of the white man; therefore, by prearrangement, wagon master Joseph Winston conveyed a number of them to Los Angeles in his returning train arriving in that city on January 4, 1861. The object was to impress the Indian with the wealth and power of the white man and to expose him to at least a representative of the "great white father." Captain Hancock was the closest thing to a "great white father" then in Los Angeles and so he entertained the Indians while they were there. Many years later Hancock's wife, Almira R. Hancock, in her book *Reminiscences of Winfield Scott Hancock,* recorded her impressions of this interesting event:

> Previous to this excitement, the Mojave Indians had been giving the troops much trouble, but had been finally subdued. After peace had been established, as a matter of policy the chiefs and other prominent Indians were invited to visit the Quartermaster in Los Angeles, as they had already been much impressed with the grandeur of his formidable train of wagons. On arrival they were formally escorted to the residence of the big chief. When passing the large market building, which had just been completed, they turned to go in, thinking it was in keeping with his great wealth, etc. Their disappointment was evident when they reached the modest little cottage, minus one gable-end, a result of the heavy rains. They gave three audible grunts, and shook their heads dubiously, that so great a man should live in so humble a manner. As soon, however, as they were taken to their tents, pitched in the corral, within a short distance of the house, and there viewed the great possessions of the Quartermaster's Department, their increased respect was apparent. Jack, a bright young Indian, who spoke a little English, and who was the heir-apparent of the Mojave chieftainship, exclaimed, "Much mulie and wagon, much pork and beanie, *house no goodie*," pointing toward the Captain's home, with evident discontent. In music they evinced much interest, and, if possible, would keep me at the piano, or "music table," so called by them, from early dawn until night. Upon one

occasion, when I was entertaining these dusky friends with "The Girl I left Behind Me," and other airs with which they were familiar, from having heard them played by drummers and fifers, at Fort Mojave, I was nearly overcome, upon completing the performance, to find half of the Indians in as complete a *deshabille* as possible to conceive, some making strenuous efforts to force their legs into the sleeves of some gaudy shirts that had been presented to them, others trying on hats, and making their toilet in numerous ways. It was time for me to make an exit, and to have the miserable creatures sent away, as they were having altogether too good a time at our expense. The principal decoration used upon these Indians' heads was mud. I have seen them sit an hour beside a puddle of water saturating their hair, and separating it in thick strands to dry. It was some time before the Quartermaster could make them understand that their visit must have an ending, and when it came they marched in Indian file through the town, decked out in his discarded dressing robes, hats and uniform, much to the merriment of those who witnessed the departure.[65]

Previously in this narrative it was noted that on August 31, 1859, when the Mohaves submitted to Armistead, they ceased to exist as an autonomous nation. They had the strength and cunning to resist after that, but they had the wisdom not to. They had been a great Nation and, it seems, among their other exemplary characteristics, they could accept defeat when it seemed to be in the best interests of their people. It saddens this writer to realize (as reflected by Mrs. Hancock's remarks and other sources of the period) that the white man had neither the insight nor the magnanimity to treat these Indians with even a basic degree of respect. They were the agents of a proud and beaten people—yet they were treated as "miserable creatures" whose curious behavior could provide "merriment" to the officers and folk of Los Angeles.

Little will be said about the Mohaves in the remainder of this narrative (although volumes could perhaps be written). It is enough to know that throughout the period covered by this paper they were still there in Mohave Valley, living an increasingly degraded existence in the presence of an increasing number of white men. In the mid-1860s many of them moved to the Colorado River Indian Reservation near Parker, Arizona. But others, however, refused to leave their ancestral lands to relocate on this reservation. After many years, these latter Indians were rewarded by the creation of a second reservation called the Fort Mojave Indian Reservation which still exists today.

THE CIVIL WAR
Prospectors On The Mojave

In early 1861, at the beginning of the Civil War, the army was forced to abandon many of its posts in Indian country. Troops were consolidated in population centers for the control of citizens, and many of the regulars were being sent back East. Fort Mojave was one of the posts thus affected. It was abandoned on May 28, 1861. The troops were marched out over the Mojave Road to the settlements of southern California. Capt. Granville O. Haller, commander of the post at that time, informed the Mohave leaders that the army would return within two years. The Indians were asked to guard the government buildings and improvements.

The troops were now gone from Fort Mojave and traffic over the road might have stopped except that a change was taking place on the desert. After the mining success at Washoe, in Nevada, prospectors were extending their explorations into other unknown desert countries. For the next four or five years—while investment in mines was the rage among the monied men of San Francisco—prospectors scrutinized the mountain ranges in every corner of the desert.

Probably the first mines worked in the eastern Mojave Desert were at Salt Springs on the Salt Lake Trail. Emigrants on that trail in 1849 noticed what they thought was rich gold-bearing rock. Then, as early as 1850, attempts were made to work the deposits. The effort was not successful in those early years because of difficulties with the Indians and because of the extreme isolation of the site.

The next major discoveries in the eastern Mojave Desert were made at Potosi, also on the Salt Lake Trail. This area was partially investigated in the late 1850s by Mormons looking for lead. It experienced a respectable boom as a silver district in 1860 and 1861.

Also, in 1861 discoveries were made in El Dorado Cañon, near the Colorado River about twenty-five miles south of where Hoover Dam is now located. This development was of importance to the Mojave Road because that route was used to reach the mines. The Mojave Road was followed to a point about six miles east of Rock Spring where there was a junction and a road angled off northeast to the vicinity of the present Searchlight, Nevada, and then on to El Dorado Cañon.

In 1862 important placer discoveries were made along the Colorado River near La Paz. Hundreds of miners rushed to this site. There was a direct road to the mines by way of San Gorgonio Pass; but, even so, some travelers used the Mojave Road to reach these new mines. They traveled to the site of then-abandoned Fort Mojave over the Mojave Road and then went down the river to La Paz.

Military activities in California reached a point early in 1863 that would permit Fort Mojave to be regarrisoned. Late in 1861 and 1862 an army had been formed in southern California and a large force—the famous Column from California—was sent over the southern route from California to New Mexico to drive the Confederates back into Texas. James Henry Carleton—at first elevated to the rank of colonel and later to brigadier-general of volunteers—commanded the expedition.

Hancock had left Los Angeles in 1861. While Carleton prepared his army, the quartermaster's depot was moved from Los Angeles to the fledgling new seaport town of Wilmington near San Pedro. Government contractor Phineas Banning donated land to the government near his new town. The Wilmington Quartermaster's Depot was built on one plot and the new army post, Drum Barracks—it was called Camp Drum for a short time—was established on another. This military complex took over the role that the Los Angeles quartermaster's depot had performed earlier. For the next ten years troops and supplies to be sent over the various roads to posts in Arizona and the desert country of California were funneled through Drum Barracks and the associated quartermaster's depot.

Late in April, 1863, Companies "B" and "I" 4th Infantry, California Volunteers, left Drum Barracks and marched out over the Mojave Road to reestablish the post among the Mohaves. They arrived at Fort Mojave on May 19, 1863. Fort Mojave would be occupied by the United States Army continuously from that time until September 29, 1890, when the site was turned over to the Indian Service. This put troops back on the Mojave Road and suddenly government supply trains, messengers, and occasional columns of soldiers utilized the road exactly as they had two years earlier. The chief difference was that the amount of civilian traffic in the form of prospectors and miners had greatly increased.

The California Volunteers stationed at Fort Mojave in the period 1863-1866 had for the most part been recruited in the mining regions of California. They had an eye for mineral wealth and immediately recognized the potential of the desert country. One of the commanders of Fort Mojave during this period—Capt. Charles Atchisson—was particularly sympathetic to this interest. He issued special passes for the men to go prospecting in the wilds of northwestern Arizona, southern Nevada, and the eastern Mojave Desert in California. The Sacramento Mining District, about thirty-five miles east and a little north of Fort Mojave, was discovered by these volunteers in 1863 and they also became active in many other mining districts and enterprises.

In 1863, a new district was formed some forty miles west of the Colorado River. It became known as the Rock Springs Mining District and was sometimes referred to as the Macedonian Mining District. This district was short-lived. The San Francisco Mining District, situated about ten miles east of Fort Mojave (in the vicinity of the present Goldroad and Oatman, Arizona), was another important development of 1863. These mines attracted many miners and capitalists from San Francisco and elsewhere. This traffic contributed to the increasing flow of military and civilian traffic on the Mojave Road.

Crossing the Colorado River had always been a problem. During the first phase of Fort Mojave occupation (1859-1861) the army had some sort of boat or skiff that was used to assist in crossing. After the fort was abandoned in 1861 the increasing number of prospectors and miners were left to their own devices to cross the great river. Then, in 1862, San Bernardino citizen John Brown established a ferry at the site of then-abandoned Fort Mojave. It seems he accurately assessed the profit to be made from the prospectors and also he probably guessed that Fort Mojave would soon be reestablished. After the fort was garrisoned once more, John Brown was given exclusive ferry rights on the military reservation for about a year. After that, the Fort Mojave ferry was usually operated by the post sutler. The major point, so far as this paper is concerned, is that after 1862 there was a ferry in operation at Fort Mojave where military and civilian parties could be crossed safely.

A second ferry was established early in 1864 at a point about six miles (by land, that is; it was farther by water) up the river from Fort Mojave. William H. Hardy established a steamboat landing and town here early in 1864. It was called Hardy's Landing for a time but soon became known as Hardyville. Hardyville existed for many years as a place of importance on the upper Colorado River. Goods destined for the interior of Arizona were shipped to Hardyville and there transferred to wagons and hauled to their destinations overland.

William H. Hardy constructed a new link in the Mojave Road. When traveling east from Pah-Ute the road to Fort Mojave would have been much out of the way if the traveler were headed for Hardyville. He would have been required to cross at Fort Mojave and travel an unnecessary six miles up the east side of the river to his destination. To remedy this, in 1864 a road was developed that left the original road at a point about seven miles from the river and headed northeast direct to Hardyville. Thus, fifteen miles after leaving Pah-Ute Creek the traveler could make his selection–Fort Mojave or Hardyville.

By the end of the year 1863, the Mojave Road was well established as a reliable route into the desert country of California and on into Arizona. Hundreds of prospectors and miners were at work in this lonely land and the Mojave Road was their highway. Meanwhile, in the mountains of central Arizona, events were taking shape that would soon throw a vast amount of traffic on the Mojave Road. This route that had been so lately proven, and had become recognized as a viable wagon route, would soon become one of the major life lines to the struggling new Territory of Arizona.

ON TO PRESCOTT

For many years it had been believed that a rich mineral country existed in the mountains of central Arizona. There were stories of Indians with golden bullets and tales of old Spanish mines. Also, mountain men told about having seen "color" in some of the streams. But the Yavapais, Apaches, and other hostile Indians had kept white men from making a thorough examination of this country.

In the spring of 1863, a group of prospectors under mountain man Joseph R. Walker made important discoveries on one of the streams not far from what is now Prescott, Arizona. It is interesting to note that several wandering Mohaves had assisted in leading Walker's party from a camp on the Gila into this new country.

The original mining district, called the "Pioneer District," was duly organized on May 10, 1863. Within weeks word was spread to the outside world by way of southern Arizona (where the prospectors had gone for supplies) that important new discoveries had been made. The rush was on. Before the end of the year there would be hundreds of men working in these mines.

Meanwhile, on February 24, 1863, President Abraham Lincoln signed into law the Act that gave territorial status to Arizona. Prior to that time, this area was included in the Territory of New Mexico.

Lincoln appointed the officials to run the government of the new territory and, after many delays, toward the end of 1863 they were on their way to Arizona. Originally, there was no question in anyone's mind where the new territorial capital would be. Tucson seemed to be the only settlement of any size. However, by the time the officials left for Arizona, things had changed. As they moved over the Santa Fe Trail toward New Mexico each arrival from the west brought additional news about fabuluous gold discoveries in central Arizona. By the time the party reached Santa Fe and Albuquerque they all had the "fever" and it was decided they would head directly for the new mines. It was felt the bulk of the citizens of the territory must be in that vicinity and that is where the capital should be. It is an interesting sidelight that by this time Gen. James H. Carleton was military commander in New Mexico and Arizona with headquarters in New Mexico. He was a strong supporter of the new mines and doubtless his enthusiasm reinforced Governor John N. Goodwin's determination to take his party to Walker's diggings.

In December the party headed west over Beale's Wagon Road into Arizona. At Navajo Springs in eastern Arizona (one of the first points the party was certain was inside the new territory) Governor Goodwin issued a proclamation announcing his new government and stating

Prescott in about 1870. *Courtesy, Arizona Historical Foundation, Hayden Library, Arizona State University; Tempe, Arizona.*

that for the moment the seat of government would be at Fort Whipple near the new gold country. From this point the party traveled on, arriving at Fort Whipple in January 1864.

Under Carleton's direction, the military had been active near the new mines since the summer of 1863. Based on reports from an expedition he had sent to the area during July-September 1863, Carleton became convinced of the future of the country. In November he dispatched a large command from New Mexico to establish Fort Whipple (the fort was named for explorer Amiel Weeks Whipple who had recently been killed at the Battle of Chancellorsville). The fort was established early in December in Chino Valley. It was originally thought the territorial capital might be at this site–and it was intended that Fort Whipple would be near the capital wherever it might be settled.

In May it was clear that the Fort Whipple site was not considered suitable for the territorial capital. Consequently, it was moved about twenty miles south to a new location on Granite Creek where the soldiers would be closer to the bulk of the miners who were then being harrassed by Indians. Meanwhile, Governor Goodwin had gone on a long trip of exploration through his territory with a view to selecting the most eligible site for the new capital. Later in May the new site of Fort Whipple was decided upon and a town was laid out near by. It was named Prescott in honor of the famous American historian William H. Prescott. His name was nominated because of the abundance of Indian ruins in this part of Arizona which were then considered by many to have been Aztec in origin. Prescott had written extensively about the Aztecs in Mexico.

A printing press had accompanied the governor's party from New Mexico and a newspaper, the *Miner*, was published at Fort Whipple and later at Prescott. This paper–as subsequent references in this narrative will show–is an important information source for Mojave Road activities during the 1860s and 70s. Regarding the location of Prescott and the routes leading to it, the *Miner* reported on June 22, 1864: "Prescott is about twelve miles west from Walker's Gulch (Lynx Creek), about six miles north from Groom's Creek, ten miles north from the Upper Hassayampa, and some thirty miles north from the town of Weaver and the Antelope diggings. From La Paz, by the Ehrenburg road to Weaver, it is not more than 150 miles; by the Williams Fork trail, not over 170 miles. From Fort Mohave it is about the same distance, while from Tucson, it is about 230 miles, a little west of north; and from the San Francisco Mountain, it is south by west nearly one hundred miles."[66]

As is implied in this article, by June of 1864 there was an established route from Fort Mojave to Prescott that provided a direct connection from California to the new capital. It is instructive for the purposes of this paper to trace the events that led to the creation of this route because it represents a 160-mile extension on the Mojave Road.

Walker's party entered the Prescott area from the south. The first soldiers to provide the miners with protection came from New Mexico. Then the governor's party, likewise, came from that direction. Also, the early supply trains to reach the area (approximately during the last six months of 1863) worked out from New Mexico. Most of this traffic used Beale's Wagon Road to the San Francisco Mountains and then turned southwest to Prescott.

This early reliance on New Mexico as a route to Arizona and as a source of soldiers and supplies was to be short-lived. It happened only because Arizona was in the military Department of New Mexico at the time and Carleton was interested in the new mines. Also, the fact that the governor's party approached the territory from the east was doubtless a factor.

It soon became apparent that Arizona was more logically supplied in all things–mail, emigrants, politicians, provisions, or soldiers–by way of California. Consequently it was not long until prospectors, freighters, and the military worked out the routes to California. There were two main routes from Prescott. One ran southwest from the capital, crossing the Colorado at La Paz, and then followed a course to San Bernardino through San Gorgonio Pass. The other route–the Mojave Road–traveled northwest from Prescott, turning more directly west and crossing the Colorado at Hardyville or Fort Mojave, following across the eastern Mojave Desert and through Cajon Pass.

The first use of the extension of the Mojave Road connecting Fort Mojave with Walker's Diggings was probably late in 1863. The report of Captain Henry M. Enos (reproduced in the appendix) shows that this route was originally known as the "Irataba trail" being named for the Mohave chief who was the first to lead emigrants from California with their wagons over the route. In December of 1863 and January of 1864 Enos, a captain in the quartermaster's department from New Mexico, conducted an exploration over the route and reported it to be practicable for wagons.[67]

By the beginning of 1864, the Mojave Road had become a vital supply line 450 miles long connecting the commercial seaport at San Pedro with Prescott. The existence of the territorial capital, the large number of miners in its vicinity, and the soldiers sent to protect them, produced a tremendous demand for supplies and services and support of all kinds from California. Soon there was a constant flow of military and civilian traffic rolling over the Mojave Road.

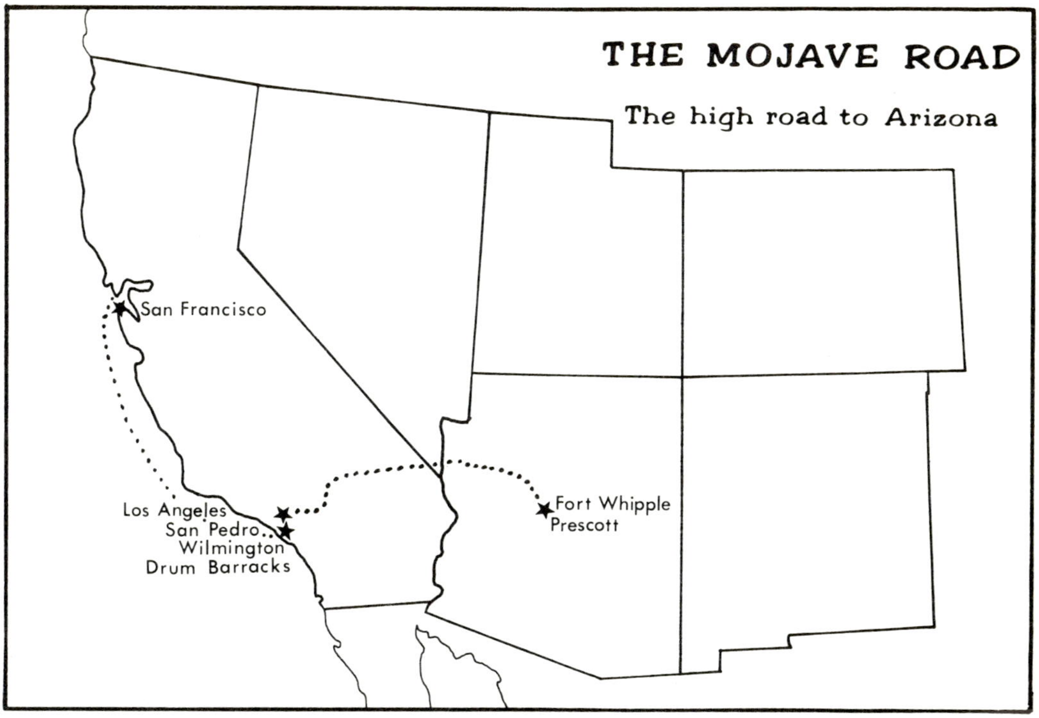

Very soon, also, Indians emerged as an important—if not the major—factor in Arizona politics and commerce. The gold mines did not immediately pan out as well as hoped. However, a large Indian population that had not been too much concerned by outsiders before, soon necessitated ever-increasing military activities in the territory. If a man could not make a living in the mines, he could engage in some activity with the military—civilians profited by providing hay and provisions, hiring out to quartermasters, serving as guides, scouts, and interpreters. Some even sold whiskey to the troops and Indians.

The first Legislative Assembly of Arizona Territory convened in Prescott in the fall of 1864. The subject of roads was very much on the minds of the legislators in this isolated wilderness capital. Several toll road companies were incorporated by Acts of the Assembly, one of which was the Mojave and Prescott Toll Road.

The franchise of this road included the 161 miles between Prescott and Fort Mojave by way of Union Pass. The original incorporators of the company included R. E. Farrington, T. M. Alexander, and others. The Act was approved on November 3, 1864. By December the wheels were set in motion for work to commence on the road. By July of the next year, the company was ready to have its road inspected by a commissioner appointed by the governor (this being in accordance with the Act) so they might begin the collection of tolls.

The commissioner discovered that some $17,361.21 had been spent on the road. More importantly to the company he stated: "In my opinion the Company has in all respects complied with the requirements of the Act incorporating said Company, and are justly entitled to collect and receive toll on said Road, according to the provisions of said Act."

On September 18, 1865, Governor John N. Goodwin issued a certificate that enabled the company to collect tolls. Toll gates were established at different times on the west end of the road in Union Pass and on the east at a point that became known as "Toll Gate," about fifty miles from Prescott. The owners were permitted to collect the following tolls: "For each wagon drawn by two horses, mules or horned cattle, 1½ cents per mile; for each carriage or cart drawn by one horse, mule or ox, 3/4 of one cent per mile; for each pack animal, ½ cent per mile; for sheep, goat or hog, 1/8 cent per mile."

William H. Hardy was not extensively involved in the toll road at first. But this changed in 1866 when the *Arizona Miner* reported "Mr. Hardy had by a large purchase become the chief owner of the road. With characteristic enterprise, he has already set men to work upon it, and he will not stop them until the road is made a first rate one in every respect." Thereafter, Hardy was recognized as the driving force behind the Mojave and Prescott Toll Road.

MAIL FOR ARIZONA

Arizona had a tremendous need for good mail service. In common with all other isolated frontier communities, Prescott and the other settlements in central and northwestern Arizona were dependent in large measure upon the mails. All manufactured goods came from California. The requisitioning, processing, and shipping of these goods required correspondence. Also, many men working in the mines in Arizona were backed by men in California. This required correspondence. And the families of many of the early pioneers were still in California in the early years. This, too, required correspondence. Add to these needs the fact that Arizonians–like Americans of all ages everywhere–longed to keep up on world news and were dependent upon California newspapers for their share, it can be seen that good mail facilities were vital to the success of the struggling new territory.

The military had a vital dependence on mail facilities. Starting in 1864, most supplies and troops for Arizona came from California. Also–and perhaps more importantly–department and division headquarters were in San Francisco. Frequently commanders in the field in Arizona had to wait for a round-trip mail to San Francisco before they could take action on some point which seemed not to be covered in their orders. At times the military made use of their own expresses to carry the mail. This technique had been used, for example, exclusively during the first phase of Fort Mojave history to carry the mail between Los Angeles and the fort.

Before leaving Washington, D. C. to journey to their territory in the wilderness, Governor Goodwin and other territorial officials urged upon the postmaster general the importance of extending existing mail facilities in New Mexico and California to include at least one line through Arizona. Then, early in 1864, when the capital was being established and the new government set in motion, petitions were circulated around Fort Whipple and Prescott asking for a mail route from Los Angeles to the new placer mines. But even if the Post Office Department was convinced that such facilities were needed, and wanted to provide them, slow-moving government machinery would cause the citizens of Arizona to wait at least a year for regular mail service.

Meanwhile, the void was filled by military and private expresses. A semimonthly military express ran from Drum Barracks to Fort Mojave. Then, the post sutler at Mojave, L. Dukes, operated a private express from that point on to Prescott. There was another private express, operated by James Grant, that connected San Bernardino and

Prescott by way of San Gorgonio Pass and La Paz. In addition to these arrangements, quantities of mail matter were carried to the territory by teamsters who happened to be going that way. Needless to say, this system was not satisfactory to the pioneers of that new territory.

In late 1864 a potential mail route was established by an Act of Congress between Los Angeles and Santa Fe by way of La Paz and Prescott. Then, a contract was awarded early in 1865 to Sanford J. Poston to carry the mails over this route between Albuquerque and San Bernardino by way of Prescott and La Paz. The mail was to be carried once every week in each direction. Service was to commence on March 1, 1865.

This put James Grant's private express on the La Paz Road out of business. Grant went to work for Poston as a subcontractor on that portion of Poston's route from La Paz to Prescott. The semi-monthly express from Drum Barracks to Fort Mojave by way of the Mojave Road was continued. However, L. Dukes–the post sutler–stopped his commercial express from Mojave to Prescott. During much of the remainder of 1865 a military express was established that connected Mojave and Fort Whipple. This maintained a mail service (although entirely run by the military) over the Mojave Road all the way from Drum Barracks to Fort Whipple.

There was an interesting difference between the two military expresses that formed this through connection from Drum Barracks to Fort Whipple. The one from Drum Barracks to Fort Mojave was carried by a civilian contractor while the one from Mojave to Fort Whipple was carried by soldiers stationed at Mojave. The explanation for this difference is that if enlisted soldiers were used for the express from Mojave to Drum Barracks, there was a good chance that when they came in sight of civilization they would be tempted to desert and not return to their isolated outpost on the Colorado. In going from Mojave to Whipple, on the other hand, there was no inducement of that sort to tempt the soldier to abandon his duty.

By early 1866 both regularly scheduled military expresses over the Mojave Road had been discontinued. After that the military at Whipple received their correspondence by way of the civilian mail route over the La Paz Road. The military at Fort Mojave were required to send an express downriver to La Paz to pick up their mail. In emergency situations the army occasionally used special expresses on the Mojave Road after this–but there was never again a regularly scheduled military express on the Mojave Road.

The contract awarded to S. J. Poston for carrying the mail between San Bernardino and Prescott by way of La Paz; and from Prescott on to Santa Fe was for the period March 1, 1865, to June

30, 1866. In late 1865 bids were requested for carrying the mail after June 30, 1866. However, there was an important change stipulated in the request for bids; it was provided that the mail between San Bernardino and Prescott would be carried by way of Hardyville instead of La Paz.

On March 29, 1866, Poston was declared the successful bidder for this modified contract to convey the mail to Prescott over the Mojave Road. It was then decided not to continue that portion of the route from Prescott to Albuquerque. This would not constitute much of a change since Poston had not carried the mail with any degree of regularity over that portion of the route anyway. The contract was made for one year to commence July 1, 1866, and the contractor was to be paid $23,247.00 for the service.

S. J. Poston had won a number of mail contracts in addition to the one over the Mojave Road. But, for some reason, he decided to get out of the business and on May 5, 1866, relinquished his rights in these contracts to James W. Parker of New York with an agreement with the Post Office Department that Parker was to perform the same services for the same remuneration.

Parker operated his enterprises from an office in New York City and later (after February 1867) from St. Louis, Missouri. There were two men who represented him on the scene. Both were veteran stage-line operators with many years of experience in California. Frank Clugage was the regional overseer. He was on the scene to help set new lines in operation or to troubleshoot when things went wrong.

The man who would run the day-to-day operation over the Mojave Road was William N. Ballard. Ballard was from Ohio, unmarried, thirty-four years of age, and had a good many years' staging experience behind him when he took over the superintendency of the mail route over the Mojave Road in 1866. He was an energetic man and a tireless traveler. The accounts of his arrivals and departures in the newspapers of San Bernardino and Prescott during 1866-1868, while the mail went over the Mojave Road, are a testimony to his constant attention to duty and to his struggle with the Indians and the elements to keep the mails going through.

The new mail was scheduled to start on July 1, 1866. Ballard would soon discover that the citizens of Arizona were in open hostility against mail contractors. Toward the end of Poston's first contract, he had become negligent. It was inevitable that Ballard and the new mail service would be watched closely and the citizens would raise a cry at the first hint of failure.

Unfortunately for Ballard, the citizens were not the only hostiles he would have to deal with. Just at the time the new mail was to commence passing over the Mojave Road, the Indians on both sides of the river were on the warpath.

The stretch of road between Camp Cady and Pah-Ute Creek was considered dangerous and had been for several years. To counter this, patrols of California Volunteers were sent from Drum Barracks to the Mojave River and Soda Lake Country several times during Civil War years. Finally, Camp Cady was reestablished by the volunteers on April 23, 1865. It was maintained by the volunteers for about a year and then after another short break it was garrisoned continuously by regulars until 1871.

In June of 1866, just weeks before the mail was scheduled to roll over the Mojave Road, the Pah-Utes in the eastern Mojave Desert committed a number of serious depredations. On June 4, 1866, at a point about ten miles east of Soda Lake, a military express rider (George Andrews) and his escort of one soldier were attacked. One horse was wounded and a horse and pack were lost. Andrews and the soldier drove the Indians off and ran for Fort Mojave, nearly one hundred miles away. They arrived there in two days with the horses exhausted and Andrews near death. The horse that had been wounded died. Andrews suffered while the post surgeon removed from his left arm a stone arrow point that he had carried for two days across a hundred miles of desert.

Then, on June 10, a civilian wagon train belonging to Daniel Hazzard was attacked near Marl Springs. A man named Leonard Taylor was killed. The wagon he was driving was in the rear of the train and was dragging a little too far behind.

But troubles with Indians on the Mojave Road had just commenced. The mining district at Macedonia Mountain ten miles northeast of Rock Spring was the next target. On June 12, 1866, a group of Indians that had formerly been friendly to the miners attacked Moses Little when he was alone in a cabin and killed him. This caused the rest of the small party of miners to abandon the district temporarily.

During this same month–June 1866–Camp Cady was regarrisoned by soldiers of the regular army. Then, on July 29, 1866, a fight took place at Camp Cady itself. A band of Indians, estimated to number thirty-six, was seen in the vicinity of the post. The officer in command, Lt. James R. Hardenbergh, 9th Infantry, assumed they were hostile and were about to attack the fort so he sallied out with six of his small force and attacked them. The soldiers were completely routed by the Indians. Three of the soldiers were killed outright and three–including Hardenbergh–were wounded. For a few days–until reinforcements arrived from Drum Barracks–it was feared the fort itself might fall to the Indians.[68]

These depredations, which were mostly successful for the Indians, doubtless increased their self-confidence and reinforced their determination to pursue a hostile course. At the same time, it raised the public cry in the settlements for troops to be brought in to protect the road.

The mail contractor would rightly be concerned about the status of the desert Pah-Utes. But an even bigger threat had been raised east of the Colorado between Hardyville and Prescott by the Walapai Indians.

The Walapais were more powerful than the Pah-Utes. They were higher on the scale of material achievement. Some of them had horses, while the Pah-Utes ate every horse they could get; also, many of the Walapais were armed with guns. The Walapai country was richer, making a better living possible. They could assemble in larger numbers and become involved in extended campaigns. If aroused, they were a much more formidable foe than the wretched Pah-Utes of the Mojave Desert.

Until 1866 the Walapais had been peaceful toward the whites. They viewed the prospecting and mining developments in the Sacramento Mining District and at other points in their country with some apprehension. Likewise, they had been somewhat concerned about the construction of the toll road between Hardyville and Prescott and the increasing number of freight wagons and troops traveling through their country. To insure their peaceful status, William H. Hardy, the proprietor of the toll road, persuaded them to enter into a treaty whereby they were to receive a specified quantity of goods from Hardy in return for which they were not to molest travelers on his road. As a result, in early 1866 it was not uncommon to find them in the camps along the road to Prescott while the trains were stopped, herding animals and assisting the heavy freight wagons up steep grades. These services were performed cheerfully and for triflings in the way of payment. They had been known to scout around for strayed stock and to return the animals to rightful owners.

It appears that the first blow in what became another major Arizona Indian war may have been struck by an Indian. The story goes that a man named Edward Clower, who was ranching at the Willows, had lost some horses and had engaged the services of a Walapai Indian to search for them. The Indian was permitted to sleep in Clower's cabin after searching for the horses. This was on the night of March 30, 1866. During the night the Indian was supposed to have killed Clower and looted the cabin, setting fire to it with Clower in it. In any event, the cabin was burned with Clower's body burned to a crisp. The cabin appeared to have been pillaged.

A few days later a wagon train, which included some of Clower's former acquaintances, came into the Walapai country headed for Hardyville from Prescott. Upon making camp on that day they were approached by Wauba-Yuma the head chief of the Walapais. Wauba-Yuma presented himself and showed the treaty he had made with Hardy. The white men consulted for a moment and decided that the Clower incident proved the Walapais were hostile and intended war anyway and proceeded to shoot the chief on the spot. This they did without giving him a hearing. A man named Sam Miller did the killing.

Resisting the temptation to analyze this case in detail, it is sufficient for this paper merely to chronicle the events to show why the white men were suddenly plunged into a war with the Walapais during the first half of 1866. It was a war that would cause large numbers of troops to be moved to Arizona to be stationed at Forts Mojave and Whipple and several intermediate points along the Mojave Road. It was a war that made Indian fighting—and support of the army—the principal business of northwestern Arizona for the next several years. It came at the time (about 1866) when interest in desert mines had subsided. Prospectors and miners in the desert country were no longer able to interest capitalists in San Francisco and elsewhere in their mines.

With the public about ready to hang anybody who identified himself as a mail contractor and with the Pah-Utes and Walapais hostile, overconfident, and seething with rage, the stage was set for the new mail to get off to an exciting start. Service was scheduled to commence on July 1, 1866, but the contractor was by no means ready to start on that day.

Clugage and Ballard were both hard at work. But it took time to get the route stocked. It was not until about August 5 that the first mail reached Hardyville from California. But by September 20, things seemed to be under control and a correspondent from Hardyville wrote "the mail now comes and goes once a week, from and to Los Angeles and is carried in light wagons."

In mid-November the prestige of the mail route on the Mojave Road advanced a considerable step with the introduction of Concord stages on the line to carry the mail and passengers. Wm. H. Hardy passed along his impression of the arrival of the first coach in Hardyville from San Bernardino: "It looked like business to see a fine four horse coach drive up to the door, and hear all hands hurrah for the mail, and some hallooed out [of] the stage, and it proved to be a regular stage with good horses."[69]

Since the very beginning, mail riders were tormented on the road between Hardyville and Prescott by Pah-Utes, Walapais, and Yavapais.

In late July, on one of his first trips, the mail rider was chased part of the way by a band of about ten Indians.

Then, on September 23, 1866, a few Pah-Utes crossed to the east side of the Colorado near Hardyville and attacked a train belonging to John Dare. They killed one teamster, a man named John Killian, and five mules.

Four days later the public was again aroused when the Walapais attacked six miners who were working in the Sacramento Mining District, some thirty miles northeast of Hardyville. Two of the men escaped with their lives although one of them was badly wounded. The other four were killed and their bodies mutilated by the Indians.

After Killian's murder, and now the massacre in the Sacramento District, the citizens around Hardyville and Mohave City became aroused and prepared to take the field against the Walapais. While making their preparations in Hardyville, they decided to drive off the Mohaves, who were in the habit of lingering around, in fear that the Mohaves might report these preparations to the Walapais. Not wishing to be placed in this false light, the Mohaves, in company with a few whites, took the field themselves. They found a village of Walapais thirty miles from Hardyville and massacred twenty-six Indians and took nineteen women and children prisoners. This completely exonerated them in the eyes of the whites around Hardyville and Mohave City.

So far, while the mail contractor may have been getting a little nervous about the depredations and wholesale murders being committed all around him, the mail had escaped unscathed. In the next adventure, however, the contractor did not fare so well.

On the evening of November 23, 1866, the mail rider with an escort of two mounted soldiers arrived at Fort Rock, which was situated on the Mojave Road between Hardyville and Prescott some ninety miles east of Hardyville. While the mail rider and his escort were saddling up the next morning they were attacked by a band of Indians which was estimated at between fifty and seventy-five, many of whom were armed with guns.

The number of whites present was only seven, which included the station keeper and his son, two teamsters, the mail rider, and the two soldiers. The defenders conducted themselves in a most creditable manner; and though three of them were wounded in the ensuing battle, which lasted for seven hours, the Indians feared to charge those deadly guns behind the breastworks of Fort Rock and they eventually withdrew. It was reported that the Indians lost between fifteen and twenty killed in the engagement. The losses for the defenders were three men wounded, eight horses and mules belonging to the mail company killed, and six head of oxen driven off.

After the battle the Indians stayed in the vicinity imposing a state of siege on the little garrison. Fortunately for the defenders, the next day a large party arrived from Prescott, lifted the siege, and assisted in removing the wounded to Hardyville.

On the road to Hardyville they met Ballard, who was on his way with light wagons and fresh stock to commence carrying the mail in wagons to connect with his stages from San Bernardino. He abandoned the project at this point and returned to Hardyville with the group. For the next few weeks it was almost impossible for the mail to get through to Prescott.

This attack on Fort Rock stopped the establishment of a through stage service from San Bernardino to Prescott just at a time when it was within a week or two of being implemented. The contractors still intended to establish such service but so long as the Indians had such a free hand they were afraid to take the risk.

Depredations along the Mojave Road were not limited to the Walapais of Arizona. The Pah-Utes of the California Desert came in for their share of public attention with a series of attacks.

On February 22, 1867, the Pah-Utes crossed the Colorado again (near Hardyville or Cottonwood Island) and killed a citizen named Thomas McCall at Union Pass. McCall was in charge of a party doing construction work on the road at that point.

About three weeks later, on March 13, 1867, the Pah-Utes struck again and in an entirely different quarter. This time they burned the buildings, corral, and fences of the station at Point of Rocks on the Mojave Road about fifty miles west of Camp Cady. The station was deserted at the time, so there were no injuries. Since there were no witnesses to the crime, it is barely possible that the deed was committed by whites, possibly even soldiers, instead of Pah-Utes. In any event, the Pah-Utes got credit for the act which put travelers and settlers on their guard.

On April 23, 1867, the Pah-Utes, who had been quiet for awhile, attacked a wagon train belonging to a man named Butler at Kingston Springs on the Salt Lake Trail. Kingston Springs was about sixty miles northeast of Camp Cady. The train was said to be loaded with machinery for the mines in Pahranagat, Nevada.

The number of Indians in the attack was estimated at sixty. The besieged made a fortification of the wagons and managed to hold the Indians at bay. The Indians were more interested in securing the mules than attacking a fortified position. Forty-two mules were driven off and none of the white men were injured. Most of the men stayed with the wagons while a few returned to San Bernardino to procure more mules and provisions.

Meanwhile, on about March 1, 1867, the mail route over the Mojave Road was changed from a weekly to a semiweekly mail, and from then on the mail passed each direction twice a week. It was carried once each week in the stagecoach. The coaches left San Bernardino on Monday. The second trip was a "muleback mail" carried with the mail loaded on the back of a mule or horse. This mail left San Bernardino on Wednesday. This put twice the number of mail parties on the road–twice the number of weak parties to tempt the Pah-Utes and Walapais.

The presence of the Arizona Overland Mail on the Mojave Road in the face of Pah-Ute, Walapai, and Yavapai hostility, required that escorts of soldiers be provided for almost every trip the mailmen made. The distances were so great that the army was forced (as was the mail contractor) to establish a number of intermediate stations along the road to provide relays for the soldiers riding escort. In this way, small outposts or camps were established at Soda Springs, Marl Springs, Rock Spring, and Pah-Ute Creek in California and at Beale's Springs, Willow Grove, Fort Rock, and at the Toll Gate on the road to Prescott in Arizona.[70]

Escort duties were performed by both infantry and cavalry. Efforts were made to make use of mounted infantry whenever possible for this duty so the cavalrymen could be saved for more active field operations. In California the escorts would be from one to six men–commonly two. In Arizona, where larger concentrations of Indians were expected, the escorts were frequently larger.

During the last six months of 1866 and the first part of 1867 very little besides protecting the mail was accomplished by the military in the deserts of southern California and northwestern Arizona. The Civil War volunteers had been ordered away to be mustered out of the service and the regulars had come in to replace them at a trickle. Depredations were being committed by the Indians for which no response would come from the military. This was particularly true at Camp Cady and Fort Mojave and elsewhere along the Mojave Road. Mail riders were attacked, miners massacred, stock killed and stolen, and about all the army did was supply small escorts for mail riders, government trains, and public officials. Citizens had to travel in large groups to provide their own protection.

But the prospects for the future were brighter. The new 8th Cavalry Regiment was being formed in northern California and the 14th and 32nd Infantry Regiments were arriving from the East. A large share of these three regiments, plus some support from the 1st Cavalry and 9th Infantry Regiments, would be available to Arizona. Now that the Civil War was no longer raging in the East, more resources of all kinds could be funneled into southern California and Arizona to take the offensive against the hostile tribes.

With the machinery in motion to provide Arizona with new soldiers and supplies to fight the Indians, Gen. Irvin McDowell, commanding the Department of California (which included Arizona), had to decide how the new program was to be administered. Earlier, he had organized all of Arizona into one district with a single commander operating in the territory. This had not worked well and it seemed the problem was coordinating operations over so large an area. Distances were too great and communications too slow to permit one man to maintain effective control.

In November of 1866 McDowell decided to divide the area that was previously the District of Arizona into five military districts: namely, District of Tucson, District of the Verde, District of Prescott, District of Upper Colorado, and District of Lower Colorado. The two districts of main interest here—and that because the Mojave Road passed through them—are the districts of Prescott and Upper Colorado.

The District of Prescott was to have its headquarters at Fort Whipple. It was to include Camp Lincoln (later Verde), Fort Whipple, the country north of the Gila River not included in the District of the Verde, and the territory to the west along the Mojave Road to a point within forty miles of the Colorado River.

The District of Upper Colorado was to include the stations on the Colorado River above Bill Williams Fork and the country on either side of the river for a distance of forty miles. Later these boundaries were extended. To the west the district embraced all of the Mojave Road between Cajon Pass and the Colorado River. The eastern limit was left somewhat unclear by the orders issued by the department, but later was construed as meaning the district included responsibility for the Mojave Road all the way from Fort Mojave to the Juniper Mountains.

Even though the orders for these new districts were issued in November of 1866, the new organization had little effect on military affairs until well into 1867. The problem was that there were few staff officers to take command of the districts and few soldiers for them to command.

During the first half of 1867, however, an officer arrived to command the District of Prescott (Col. John Irvin Gregg, 8th Cavalry, passed over the Mojave Road to his new district in March) and another officer for the District of Upper Colorado (Maj. William Redwood Price, 8th Cavalry, arrived at Fort Mojave on June 14, 1867). Also, throughout 1867, hundreds of new troops and long trains of army supplies moved over the Mojave Road into Arizona. It is quite likely that 1867 was the busiest year the Mojave Road ever had.

Meanwhile, hostilities continued on both sides of the river. In June skirmishes took place between the mail parties and Pah-Utes at Marl Springs and at a dry camp along the road between Pah-Ute Hill and Rock Spring. In Arizona, troubles took place with the Walapais and Yavapais at many points on the road to Prescott. On June 26 a mail carrier was killed near Beale's Springs.

In June of 1867, when Major Price first arrived and took command of Fort Mojave and the District of Upper Colorado, there were only two small companies at the post, Company "E" 14th Infantry and Company "K" 8th Cavalry. These two companies amounted to about one hundred men. Also, in his district but not at Mojave, there was Company "K" 14th Infantry at Camps Cady and Rock Spring and Company "D" 9th Infantry at Camp El Dorado about fifty miles up the river from Mojave.

At this time the Walapais had become increasingly hostile. Price was convinced that defeating them would require the coordinated movement of a large force directed against them in four or five detachments simultaneously. His effective force at Mojave was then too small to permit any immediate plan of chastising the Walapais. For the time being he had to be satisfied to concentrate his forces at Mojave and prepare his men and supplies for an extensive coordinated campaign.

To augment his force at Mojave, Price obtained permission to reduce Camp El Dorado to a small outpost and to move most of Company "D" 9th Infantry from that post to Mojave. Also, he had convinced department headquarters that a contingent of troops destined for the District of the Verde should be detained temporarily in the District of Upper Colorado for operations against the Walapais.

These activities gave Price quite a respectable number of troops to utilize against the Walapais. By the end of summer he had all of Companies "K" and "L" of the 8th Cavalry (about 160 men), Company "D" 9th Infantry (about 100 men), Company "E" 14th Infantry (about 70 men), and a detachment of Company "E" 1st Cavalry (about 20 men). This total of about 350 men, of which 180 were cavalry, were all well supplied.

In late August Price sent a detachment, including most of Company "E" 14th Infantry, to Willow Grove on the road to Prescott to establish a new army post, thereafter known as Camp Willow Grove. The purpose of the post was to serve as a depot of supplies for the Walapai Campaign.

At this point the Walapais executed another massacre, giving new impetus to the program. The massacre occurred in the Sacramento Mining District some thirty or forty miles from Mojave on August 20. Four men of the well-known Willing Exploring Party were killed.

Beginning on September 25, 1867, Price's army, christened the "Army of the Colorado" by a civilian, took the field. They would not come back to Mojave until December. They would hunt the elusive Walapai, and they would find him.

For more than two months, Price's troops scoured the desert and mountains of northwestern Arizona. They had some successes—surprising rancherias and killing a number of Indians. And, in some cases where the Indians detected the approach of the troops and were able to flee in time, the troops destroyed their homes and provisions. Still, it did not bring the Walapais to their knees and they continued to be hostile for more than a year.

Meanwhile, while Price was absent from Mojave hunting Walapais, the Pah-Utes struck again on the California side of the river. On October 16, 1867, the mail buggy (in June the Concord stages had been replaced with small buggies) had stopped at the Caves in Cave Canyon. The party consisted of the mailman Sam Button, Dr. Merrill E. Shaw, an army surgeon who was on his way to Fort Mojave, and Pvt. Timothy Donovan, the sole soldier of Company "K" 14th Infantry from Camp Cady that had been detailed to escort this mail.

After taking lunch at the Caves the party started on in the direction of Soda Springs. At a point not far from the Caves, the party was attacked without warning by about fifteen or twenty Indians. At the first fire, which was reported to be "half a dozen bullets & arrows, delivered at not more than ten yards," Dr. Shaw was hit in the neck and the escort's mule was struck in the shoulder. Donovan abandoned the mule and jumped on the mail buggy. Button whipped his mules and headed for Soda Springs, some fourteen miles distant, while Donovan cut the baggage loose, which included the mail and Shaw's trunk.

The Indians chased the buggy for some eight or ten miles. These Indians had no horses and so, in spite of the fact that the two mules pulling the buggy had already traveled almost twenty miles that day and were now pulling through heavy sand, they could keep ahead of the Indians.

Arriving at Soda Springs the party met a large army column under Capt. James P. Brownlow, 8th Cavalry, *en route* to Arizona. This ended any plan the Indians might have had to follow up their earlier attack. Dr. Shaw died at Soda Springs the next day and was buried there.

Brownlow's train left Soda Springs late in the afternoon of October 17, reaching Marl Springs the next morning. Upon arriving there he was surprised to learn the station at the springs had been under a state of siege since the night before. At dusk the evening before a band of Indians attacked the outpost and laid siege to the place all night, being driven away only by the approach of Brownlow's command. There were no casualties but, needless to say, the men were relieved to see Brownlow's command come into view.

From all appearances, things were going from bad to worse with the Pah-Utes; but, exactly the opposite was true. Major Price was busy working on the Pah-Utes through diplomatic strategy rather than force of arms. When he took his Army of the Colorado into the field in September he had taken two Indians along, at least one of which was a prominent Pah-Ute. He was assisted in making this fortunate arrangement by civilian James H. Ferry who lived in Callville and who was on good terms with the Pah-Utes. Price's idea was to create hostile feelings between the Pah-Utes and Walapais and to impress the Pah-Utes with the power of the white man.

Price kept these two Indians with him for about a month. They saw the soldiers emerge victorious from every skirmish with Indians and they saw them destroy the Walapais' homes and provisions. On October 29 Price concluded his guests had probably seen enough to achieve his ends so he sent them back to Mojave. It was agreed between Price and the Pah-Utes that they would get all the Pah-Utes to come into Mojave for a big talk.

On November 20, while at Camp Willow Grove, Price received a message from Fort Mojave with the somewhat startling news that some sixty well-armed Pah-Ute warriors had arrived at Mojave prepared to have the "big talk." The next day Price took twenty-five men of Company "K" 8th Cavalry and rode into Fort Mojave, arriving there late at night. Early the following morning, November 22, preparations were made for the conference.

Price proposed to the Indians that they move to the Colorado River Indian Reservation down the river near La Paz. They did not want to leave their ancestral lands to do this. Also, they did not believe they could live in close proximity to the Mohaves (many of whom had moved to that reservation) there as the two tribes were enemies. As a compromise, Price persuaded them to leave several hostages at Fort Mojave as insurance for the good conduct of the various bands. Price was to feed and provide for these hostages as long as they remained at the post. Also, the Indians agreed to have no communication with the hostile tribes of Indians with whom they had dealt in the past, particularly the Walapais. It was felt that the Walapais obtained some of their munitions from the Pah-Utes who in turn procured them from Mormons. They agreed to provide Price with any information they might acquire from time to time about the operations and movements of these hostile Indians. Price gave papers to the captains of the various bands that attested to these arrangements.

These agreements of November 22, 1867, seem to have settled the Pah-Ute problem along the Mojave Road in California. To my knowledge there were no more depredations committed in this country after the meeting with Price.

Price's troops came in out of the field (some to Fort Mojave and some to Camp Willow Grove) early in December. By the end of that month it was apparent to Price and everyone else that the Walapais were not beaten. As soon as the troops were back in their forts, the Indians resumed the offensive, attacking the mail parties and the small stations along the Mojave Road. Price began preparations for follow-up campaigns against these persistent Indians. Meanwhile, traffic on the Mojave Road, particularly in the California stretch, slowed almost to a halt late in 1867 and early in 1868 because heavy rains had rendered the roads all but impassable.

At about this time, an extensive railroad surveying party arrived from the east. This party had been sent into the field by the Union Pacific Railroad, Eastern Division, and was commanded by William Jackson Palmer, late a general in the volunteer army.

Congress had already decided that the first transcontinental railroad would be built much farther north, running west from Omaha to Cheyenne and on through Salt Lake City to Sacramento and San Francisco. The road was to be built simultaneously from California by the Central Pacific Railroad and from the eastern end by the Union Pacific Railroad. In addition to these arrangements, Congress had provided for a number of feeder lines which would enter this main line east of the Rocky Mountains. In the beginning the Union Pacific Railroad, Eastern Division, was one of these feeder lines.

However, while the first transcontinental road was under construction and work was commenced also on the feeder lines, the U.P.R.R.E.D. fell into the hands of men who attempted to turn it into a second transcontinental road instead of just a feeder line. In connection with their promotional schemes, they sent extensive surveying parties out into the field in 1867 and 1868 to attempt to locate a practicable route along either the 32nd or 35th parallels of latitude. They were certain of finding such a route, which they ultimately did along the 35th parallel, and then they would use the results of their explorations and surveys to attempt to persuade Congress to grant them the necessary aid and subsidies to build the road. They were, however, unable to push their measures through Congress and in 1869 the whole deal fell through with the name of the railroad being changed to the Kansas Pacific. But early in 1868, when General Palmer passed through northwestern Arizona and the eastern Mojave Desert driving his line of stakes into the sand, all the members of the party were full of confidence that they would be building a railroad along this route within the next year or two.

In 1854 when Lieut. Amiel W. Whipple had surveyed a line for a railroad through the eastern Mojave Desert, he had been in a hurry by the time he reached this point because of shortages of supplies. Instead of exploring for the best route through the general latitude that he must traverse, he took the route the Mohave Indians were accustomed to use. This was the line that later became the Mojave Road. But, because of the high mountains that cross this path, it was by no means the best line for a railroad. A line farther south would avoid these mountains and result in a level road that would be much cheaper to construct originally, and cheaper to maintain and operate once completed.

General Palmer had more time to explore and was not in the least hampered by that great impediment to most desert exploration—water. If anything, because of the great rain storms that were occurring at that time, he had too much water. He and his engineers conducted a thorough survey of the eastern Mojave and discovered the natural passage for a railroad lying some twenty or thirty miles south of the line of the Mojave Road. They passed through the desert country in January and February of 1868, crossed the Mojave River above Camp Cady, ran their line through the Tehachapi Pass, and on to San Francisco. The line they staked out between the Colorado and the Mojave Rivers was very close to the line that was utilized more than ten years later for the tracks of the Southern Pacific (later Atlantic & Pacific and still later Santa Fe) Railroad. Additionally, they staked out another line designed to connect the 35th parallel with San Diego. This route headed southwest near the present Amboy and made use of Morongo and San Gorgonio Passes.

At about this same time it became apparent that the mail route would be changed back to the La Paz Road. The exact reasons for this change are not known. During the early part of 1867 a postal inspector, Benjamin C. Truman, had toured Arizona inspecting operations there. At the conclusion of his tour he was reported by the San Bernardino *Guardian* as being strongly in favor of the La Paz Road as the mail route to Arizona.

San Bernardino *Guardian* editorials suggest the citizens of that town also favored the La Paz route. This is not surprising since traffic to Arizona over the La Paz Road passed through San Bernardino, whereas much of the traffic for the Mojave Road that originated in Los Angeles or San Pedro, took a cutoff at Cucamonga Rancho direct to the mouth of Cajon Canyon thereby bypassing San Bernardino.

It is a little surprising the change was made to the Mojave Road in the first place. The following article from the Prescott newspaper published in November of 1867 sheds some light on that aspect of the subject:

> The people of Wickenburg and La Paz do not ask for the establishment of a new route, but they want and must have service upon their old route, which was taken off and discontinued through the meddling advice of some political trickster at Washington to further some scheme of himself and friends.[71]

An editorial in the San Bernardino *Guardian*, suggests that the "political trickster" was Arizona's "late delegate to Congress" (the reference is probably to John N. Goodwin). The article goes on to state that the "late delegate" had such a grudge against the people of La Paz that he represented to officials in Washington that the town was deserted. Hence, the route was changed to the Mojave Road in 1866.

An important factor in changing the mail back to the La Paz Road in 1868 was the great losses that had been sustained by the contractor at the hands of the Indians on this route. One writer estimated the contractor had lost almost one hundred head of stock to the Indians in one year. Another source says the contractor had lost more than fifty head of mules and two of his men had been killed at the stations. Either way, these losses took large chunks out of the contractor's expected profits and prevented him from establishing through stage service on the route.

By March the contractor had his arrangements perfected with the Post Office Department and it was announced in San Bernardino that the route change would be made as soon as the stock could be moved from one route to the other and stations built on the La Paz Road. Shortly thereafter, and almost as if they knew they would not have many more opportunities to attack the small mail parties, the Walapais struck at the mailman one last time on the Mojave Road.

This attack took place at a point about four miles west of Camp Willow Grove. The mail was being carried by Charles Spencer. He was accompanied by an escort of two men of Company "E" 14th Infantry.

The picture shown on the opposite page was taken at Fort Mojave in early 1868. The Indians are Mohaves. The army officer is **Maj. William Redwood Price** of the 8th Cavalry. Price commanded Fort Mojave and the Military District of Upper Colorado in the years 1867-1869 and became importantly involved in affairs along the Mojave Road all the way from Cajon Pass nearly to Prescott. *Courtesy, The Boston Public Library.* ➡

At the first fire both soldiers were killed. Spencer grabbed his weapons and ammunition and sought cover. The Indians forgot him for the moment, occupying themselves with stripping and mutilating the bodies of the slain soldiers. This Spencer could not tolerate to witness, so he opened fire on the Indians with his repeating rifle, killing several. The Indians attacked him and succeeded in wounding him; but he took refuge among some rocks which provided an excellent fort.

For the rest of the afternoon the Indians laid siege to Spencer, rushing him several times, but he held his ground. After several hours the shooting was heard at Camp Willow Grove. Some soldiers came out to investigate and eventually Spencer was rescued.

And thus ended the carrying of the mail over the Mojave Road in the 1860s. The Indians would have no more mail riders to ambush. Shortly after the attack the shift was made to the new route. The first mail over the La Paz route arrived in Prescott on March 26. By April 18 the switch was completed and the Prescott newspaper was able to state that the mail "comes and goes with great regularity" over the new route.

After the mail was withdrawn from the Mojave Road, military posts were maintained for some years at Camp Cady, Fort Mojave, Camp Hualpai (or Toll Gate), and Fort Whipple. But the other little outposts that had been established were all abandoned.

Major William R. Price continued his Walapai Wars until mid-1869 when hostilities slowed down. The Walapais were estimated to number fifteen hundred souls. On May 30, 1868, Price wrote that the Walapais now desired peace and that the troops under his command had killed more than one hundred of them over the past two years.

THE PRODUCTIVE YEARS
1870–1883

The 1870s ushered in a period of renewed interest in mining. Before the decade was over many important developments were made in the country served by the Mojave Road in California and Arizona. Mining communities sprang up—like Ivanpah in California and Cerbat and Mineral Park in Arizona—and supported large populations in the desert country. Large mills were put in operation and some of the mines were made productive.

Military operations along the Mojave Road in the early 1870s continued to contribute to its traffic although, by this time, steamboat transportation on the Colorado had become better developed. Many supplies, even companies of troops, were conveyed to the territory over that route. By the end of the 1870s it was an uncommon thing for the military to send traffic to California by way of the Mojave Road. In one instance, an officer at Fort Mojave had to hire a guide to accompany a wagon he wanted to send to San Bernardino for vegetables. No one at the post knew the way.

The Indians of the desert country were no longer a threat. Many had been killed or were on reservations. Some went to work for white men around their mines and ranches. In many cases they pursued a fragmented and degraded form of their original ways.

More homesteads were filed upon along the Mojave River and with the increased white population and reduced threat posed by Indians, Camp Cady was no longer needed. It was abandoned on April 24, 1871. It eventually became a ranch and is known as the Camp Cady Ranch today.

Camp Hualpai, that had been established at the Toll Gate on the road to Prescott on May 9, 1869, was abandoned by the military on July 31, 1873. It had served as an operations point against the Walapais and Yavapais.

An army post was established at Beale's Springs (called Camp Beale's Springs) on March 25, 1871. The post was maintained there as a center to concentrate the Walapai Indians. They were issued rations and in that way kept off the warpath. On April 6, 1874 the post was abandoned. The Indians had been forced to move to the Colorado River Indian Reservation near Parker. Later they returned to their homeland and today they have a large reservation of their own in northwestern Arizona.

Fort Mojave continued standing guard among the numerous Mohaves until September 29, 1890, when the reservation was turned over to the Indian Service.

Army wives followed their husbands to frontier military posts even in the early days. This classic family portrait was taken at Fort Mojave in 1868. It features Captain (later Lieutenant-General and first Chief of Staff of the U. S. Army) **Samuel B. M. Young**, 8th Cavalry, his wife Margaret, their two daughters Edith and Burton, and a Mohave Indian girl who became part of the Young household while they were at Mojave. Copies of this interesting picture have been obtained from the collections of the following: *Boston Public Library, Missouri Historical Society, and Alice Margaret Young Knight–granddaughter of General Young.*

Fort Whipple not only endured through the 1870s, but it was to become the longest-lasting of all the military posts along the Mojave Road. It was not abandoned until February 14, 1913.

Prospectors and miners kept the level of traffic high on all segments of the Mojave Road in the 1870s. Then, in 1875, the California and Arizona Stage Company obtained a mail contract to run to Arizona over the Mojave Road. Once again there was regular stage service over this route. But this time there was no need for escorts and all the little military posts were gone. The following announcement appeared in many numbers of the San Bernardino *Guardian*:

> Superior inducements are now offered to persons visiting Arizona and New Mexico, by the California and Arizona Stage Company, to travel by their new route via Cajon Pass, Mojave River, Soda Lake, Hardyville, Mineral Park, Cerebat, McCrachen Mines and Williamson Valley, to Prescott, as the journey is made in the pleasantist possible manner. Good horses, skillful drivers and good comfortable coaches. Stages leave San Bernardino for all the above named points Wednesday's and Saturday's of each week.[72]

Once again regular stages ran over the route. There were stations at some of the springs across the desert.

In the 1870s immense herds of sheep and other stock were driven over the Mojave Road to Arizona and some on to New Mexico. These migrations were partly due to the discovery of new ranges in Arizona that were becoming safe from the Indian threat. At other times, large herds were driven from California during periods of drouth. During 1875 alone, more than thirty thousand sheep (and perhaps as many as fifty thousand) were crossed over the river into Arizona at Hardyville. The movement of livestock over the Mojave Road was a major business throughout the decade.

In April of 1880, after a lapse of many years, the Mojave Road suddenly acquired a military purpose once again. There had been a few murders of white men on the Colorado River Indian Reservation and, as a result, the Chemehuevi Indians had taken to the desert mountains in California and were considered for the moment to be on the warpath. The army quickly dispatched units into the desert to intercept the Chemehuevis before they could join other Indians in the north and perhaps trigger a general uprising. Two companies were sent out on the Mojave Road to patrol between Soda Lake and Pah-Ute Creek. The companies involved were "I" 1st Cavalry and "H" 6th Cavalry under Captains Camillo C. C. Carr and William M. Wallace, respectively. Carr covered the stretch of road between Soda Lake and Marl Springs and Wallace patrolled between Pah-Ute and Marl Springs.

Two civilians who showed Carr and his company the way to Soda Lake from San Bernardino, later reported to the newspaper in that town: "The company made two charges on the road, at one time hitting two rattlesnakes, and the other capturing a desert turtle. Not an unfriendly Indian was met or heard of." At Marl Springs the soldiers found: "three Piute bucks and one squaw...working for the station keeper there."

By mid-May the affair with the Chemehuevis was peacefully settled. The soldiers were no longer needed on the Mojave Road. Company "I" 1st Cavalry returned toward San Bernardino from whence they had come and Company "H" 6th Cavalry returned to Arizona.

As the 1880s dawned, the end was in sight for the Mojave Road as a major trans-desert wagon road. The long-awaited railroad was coming to the 35th parallel. The trains could not climb through the rugged Pah-Ute Mountains nor was it sensible for them to follow a grade up, over five thousand feet in passing through the Providence Mountains, when twenty miles or so to the south, a much lower and more level grade was available.

On August 9, 1883, the last rail was laid near Needles, California, connecting the line the Southern Pacific R. R. had built from the west with that built by the Atlantic and Pacific R. R. from the east. The Atlantic and Pacific soon took over the Southern Pacific tracks across the eastern Mojave and, of course, this entire line is the Santa Fe Railroad of today.

At long last the dream of Garces, Whipple, Beale, and a host of others was realized. The great avenue of commerce along the 35th parallel was a reality.

As has been mentioned, the route of the railroad was off-set twenty miles or so from the trace of the old Mojave Road. Later trans-desert wagon roads and automobile highways evolved from the track of the railroad and not from the trace of the earlier wagon road. Travel through the desert country tended to be along the railroad and then north and south from the stations along the tracks.

This has had a happy result for us. A 130-mile stretch of the old wagon road, from Camp Cady on the west eastward to the crossing of the Colorado at Hardyville and Fort Mojave, has been left in nearly original wagon road form. There are a few places where it is crossed by power lines, highways, and even a railroad, but for the most part the country is in the same form today that it was back in 1883 when the railroad provided it with such stiff competition. And, throughout most of this distance the original trace of the old road can be found.

APPENDIX

The principal objective of the *Tales of the Mojave Road* series is to increase awareness and appreciation of this historic route of travel. The books are designed to accomplish this by presenting source material pertaining to the general subject and telling the stories that go along with these materials. Some of this source material can be sandwiched into the narrative part of the book. Other important sources cannot be included in this way because of length or level of detail. This appendix is provided to accommodate a few documents in this latter category.

A word of explanation about the selection of documents to appear in this appendix is in order. In making these selections the *Tales of the Mojave Road* series as a whole is considered. For example, an extract from the Sitgreaves report is included here because of its pertinency and also because no other books in the series in the near future are planned to deal with Sitgreaves. On the other hand, I have quoted very little in this book pertaining to the Whipple expedition (even though there is a wealth of documentation available) because of plans to treat that expedition separately in the future. Also, there are no quotations in the present work from the most active military period along the Mojave Road (say, 1866-1868) because much of this period has been treated in the first four books in the series. Camp Cady and Fort Mojave have yet to be treated in detail but this is to be done in subsequent books in the series.

Extract from the report of Lt. John Charles Fremont, during April 1844 covering his approach to the Old Spanish Trail near Cajon Pass and his journey along that trail down the Mojave River.

Our road the next day was still in an easterly direction along the ridge, over very bad travelling ground, broken and confounded with crippled trees and shrubs; and, after a difficult march of 18 miles, a general shout announced that we had struck the great object of our search—*The Spanish Trail*—which here [near Cajon Pass] was running directly north. The road itself, and its course, were equally happy discoveries to us. Since the middle of December we had continually been forced south by mountains and by deserts, and now would have to make six degrees of *northing*, to regain the latitude on which we wished to cross the Rocky mountains. The course of the road, therefore, was what we wanted; and, once more, we felt like going homewards. A *road* to travel on, and the *right* course to go, were joyful consolations to us; and our animals enjoyed the beaten track like ourselves. Relieved from the rocks and brush, our wild mules started off at a rapid rate, and in 15 miles we reached a considerable river [the Mojave River], timbered with cottonwood and willow, where we found a bottom of tolerable grass. As the animals had suffered a great deal in the last few days, I remained here all next day, to allow them the necessary repose; and it was now necessary, at every favorable place, to make a little halt. Between us and the Colorado river we were aware that the country was extremely poor in grass, and scarce for water, there being many *jornadas*, (days' journey,) or long stretches of 40 to 60 miles, without water, where the road was marked by bones of animals.

Although in California we had met with people who had passed over this trail, we had been able to obtain no correct information about it; and the greater part of what we had heard was found to be only a tissue of falsehoods. The rivers that we found on it were never mentioned, and others, particularly described in name and locality, were subsequently seen in another part of the country. It was described as a tolerably good sandy road, with so little rock as scarcely to require the animals to be shod; and we found it the roughest and rockiest road we had ever seen in the country, and which nearly destroyed our band of fine mules and horses. Many animals are destroyed on it every year by a disease called the foot evil; and a traveller should never venture on it without having his animals well shod, and also carrying extra shoes.

The morning of the 22d was clear and bright, and a snowy peak to the southward shone out high and sharply defined. As has been usual since we crossed the mountains and descended into the hot plains, we had a gale of wind. We travelled down the right bank of the stream, over sands which are somewhat loose, and have no verdure, but are occupied by various shrubs. A clear bold stream, 60 feet wide, and several feet deep, had a strange appearance, running between perfectly naked banks of sand. The eye, however, is somewhat relieved by willows, and the beautiful green of the sweet cottonwoods with which it is well wooded. As we followed along its course, the river, instead of growing constantly larger, gradually dwindled away, as it was absorbed by the sand. We were now careful to take the old camping places of the annual Santa Fe caravans, which, luckily for us, had not yet made their yearly passage. A drove of several thousand horses and mules would entirely have swept away the scanty grass at the watering places, and we should have been obliged to leave the road to obtain subsistence for our animals. After riding 20 miles in a northeasterly direction, we found an old encampment, where we halted.

April 23.–The trail followed still along the river, which, in the course of the morning, entirely disappeared. We continued along the dry bed, in which, after an interval of about 16 miles, the water reappeared in some low places, well timbered with cottonwood and willow, where was another of the customary camping grounds. Here a party of six Indians came into camp, poor and hungry, and quite in keeping with the character of the country. Their arms were bows of unusual length, and each had a large gourd, strengthened with meshes of cord, in which he carried water. They proved to be the Mohahve Indians mentioned by our recent guide; and from one of them, who spoke Spanish fluently, I obtained some interesting information, which I would be glad to introduce here. An account of the people inhabiting this region would undoubtedly possess interest for the civilized world. Our journey homeward was fruitful in incident; and the country through which we travelled, although a desert, afforded much to excite the curiosity of the botanist; but limited time, and the rapidly advancing season for active operations, oblige me to omit all extended descriptions, and hurry briefly to the conclusion of this report.

The Indian who spoke Spanish had been educated for a number of years at one of the Spanish missions, and, at the breaking up of those establishments, had returned to the mountains, where he had been found by a party of *Mohahve* (sometimes called *Amuchaba*) Indians, among whom he had ever since resided.

He spoke of the leader of the present party as "*mi amo*," (my master.) He said they lived upon a large river in the southeast, which the "soldiers called the Rio Colorado;" but that, formerly, a portion of them lived upon this river, and among the mountains which had bounded the river valley to the northward during the day, and that here along the river they had raised various kinds of melons. They sometimes came over to trade with the Indians of the Sierra, bringing with them blankets and goods manufactured by the Monquis and other Colorado Indians. They rarely carried home horses, on account of the difficulty of getting them across the desert, and of guarding them afterwards from the Pa-utah Indians, who inhabit the Sierra, at the head of the *Rio Virgen,* (river of the Virgin.)

He informed us that, a short distance below, this river finally disappeared. The two different portions in which water is found had received from the priests two different names; and subsequently I heard it called by the Spaniards the *Rio de las Animas,* but on the map we have called it the *Mohahve* river.

April 24.—We continued down the stream (or rather its bed) for about eight miles, where there was water still in several holes [from later comments by Fremont it would appear that this was the site where Camp Cady would be established in 1860], and encamped. The caravans sometimes continue below, to the end of the river, from which there is a very long *jornada* of perhaps sixty miles, without water. Here a singular and new species of acacia, with spiral pods or seed vessels, made its first appearance; becoming henceforward, for a considerable distance, a characteristic tree. It was here comparatively large, being about 20 feet in height, with a full and spreading top, the lower branches declining towards the ground. It afterwards occurred of smaller size, frequently in groves, and is very fragrant. It has been called by Dr. Torrey *spirolobium odoratum.* The zygophyllaceous shrub had been constantly characteristic of the plains along the river; and here, among many new plants, a new and very remarkable species of eriogonum (*eriogonum inflatum,* Torr. & Frem.) made its first appearance.

Our cattle had become so tired and poor by this fatiguing travelling, that three of them were killed here, and the meat dried. The Indians had now an occasion for a great feast, and were occupied the remainder of the day and all the night in cooking and eating. There was no part of the animal for which they did not find some use, except the bones. In the afternoon we were surprised by the sudden appearance in the camp of two Mexicans–a man and a boy. The name of the man was *Andreas Fuentes*; and that of the boy, (a handsome lad, 11 years old,) *Pablo Hernandez.* They belonged to a party consisting of six persons, the remaining four being the wife of Fuentes, the father and mother of Pablo, and Santiago Giacome, a resident of New Mexico. With a cavalcade of about thirty horses, they had come out from Puebla de los Angeles, near the coast, under the guidance of Giacome, in advance of the great caravan, in order to travel more at leisure, and obtain better grass. Having advanced as far into the desert as was considered consistent with their safety, they halted at the *Archilette* [later called Resting Springs], one of the customary camping grounds, about 80 miles from our encampment, where there is a spring of good water, with sufficient grass; and concluded to await there the arrival of the great caravan. Several Indians were soon discovered lurking about the camp, who, in a day or two after, came in, and, after behaving in a very friendly manner, took their leave, without awakening any suspicions. Their deportment begat a security which proved fatal. In a few days afterwards, suddenly a party of about one hundred Indians appeared in sight, advancing towards the camp. It was too late, or they seemed not to have presence of mind to take proper measures of safety; and the Indians charged down into their camp, shouting as they advanced, and discharging flights of arrows. Pablo and Fuentes were on horse guard at the time, and mounted, according to the custom of the country. One of the principal objects of the Indians was to get possession of the horses, and part of them immediately surrounded the band; but, in obedience to the shouts of Giacome, Fuentes drove the animals over and through the assailants, in spite of their arrows; and, abandoning the rest to their fate, carried them off at speed across the plain. Knowing that they would be pursued by the Indians, without making any halt except to shift their saddles to other horses, they drove them on for about sixty miles, and this morning left them at a watering place on the trail, called Auga de Tomaso [better known as Bitter Springs]. Without giving themselves any time for rest, they hurried on, hoping to meet the Spanish caravan, when they discovered my camp. I received them kindly, taking them into my own mess, and promised them such aid as circumstances might put it in my power to give.

April 25.—We left the river abruptly, and, turning to the north, regained in a few miles the main trail, (which had left the river sooner than ourselves,) and continued our way across a lower ridge of the mountain, through a miserable tract of sand and gravel. ... ...

Extract from the report of Lt. Lorenzo Sitgreaves in November 1851 when he was among the Mohave Indians.

November 7, [1851], Camp No. 33.—A well-worn trail leads down the river, by the side of which in several places were found traced on the ground Indian heiroglyphics, which Mr. Leroux and a Mexican of the party, who had passed many years among the Comanches, interpreted into warnings to us to turn back, and threats against our penetrating farther into the country. We had not gone far before Indians were seen in front in considerable numbers, who appeared to be assembling to dispute our advance. By the exchange of friendly signs, three of them, mounted on fine horses, were induced to approach, whom a few presents sufficed to convince of our peaceful intentions; and they joined the party, and accompanied its march. As we proceeded their number received accessions at every step, until it amounted to some two hundred men, women, and children, who followed on foot, running by the side of the mules, and talking and laughing with every appearance of friendship. In the evening the camp was crowded with them, bringing in for barter small quantities of pumpkins, beans, corn, and, in one or two instances, of wheat, which seem to be the staples of their food, for no animals, except a few horses, were seen among them; and the few sheep we had left were the objects of great admiration, especially to the women.

The appearance of the Mohaves is striking, from their unusual stature, the men averaging at least six feet in height; and their stalwart and athletic figures offered a convincing proof of the excellence of a vegetable diet. Almost all the men were naked, with the exception of the breech-cloth. The hair, cut square across the brows in front, hung in loose braids behind, reaching frequently as low as the waist; occasionally it was matted on the top of the head into a compact mass with mud, for the purpose of destroying the vermin that infest them. The only garmet worn by the women was a long fringe of strips of willowbark wound around the waist, and falling as low as the knees. No covering to the feet was worn by either sex. Their arms are the bow and arrow, the spear and the club. The arrow is formed of two pieces—that to which the barb is attached, of hard wood, seven inches long, or one-fourth the entire length; and the other of a light reed that grows profusely along the banks of the river, feathered, as usual, at the extremity. The custom still prevails among them of carrying a firebrand in the hand in cold weather, which is mentioned in the account of Coronado's expedition in 1540, and induced those discoverers to give to the river the name of Rio del Tizon. Their lodges are rectangular, formed of upright posts imbedded in the ground, and rudely thatched on the top and three sides; a portion of the interior altitude being sometimes obtained by excavation. I saw none of so great a size as those described in the account just referred to.

November 8, Camp No. 34.—A large crowd of men, women, and children continued to follow us, many of them carrying beans and pumpkins, and all urgent for us to encamp among them, for the purpose, as they gave us to understand, of trading. I was myself anxious to obtain supplies from them; but their numbers and importunity had been so troublesome the day before, that it was resolved to exclude them from the camp, and to adopt some plan which should free us from a repetition of the annoyance. Before unpacking the mules, therefore, a chain of sentinels was placed around them, with instructions to prevent the entrance of the Indians, and places were designated on the outside where they might hold their market. This arrangement gave great dissatisfaction, and did not fully answer the purpose intended; for many eluded the vigilance of the sentinels, or took advantage of their negligence, and the camp was soon again filled with them. A large number were observed to have arms; and the fact that no chiefs had presented themselves, notwithstanding our frequent demands for them, was regarded as suspicious, and calling for all possible vigilance. The retreat was therefore sounded, and the Indians ejected from camp, which was accomplished with difficulty, and hardly without the use of violence. They left us with scowling faces, and some old women were vociferous with what we supposed to be their threats and denunciations.

November 9, Camp No. 35.—While preparing for our departure before daylight, Dr. Woodhouse, who was warming himself by the fire, received an arrow through the leg, fortunately without doing him much injury. Several others were thrown into the camp and among the mules, but the darkness caused them to fall harmless. The sentinels, however, were thrown farther out, and we got under way without further annoyance, numbers following us with yells of defiance, but taking care to keep at a respectful distance.

Extract from the report of Lt. Robert S. Williamson in November 1853 when conducting his survey of the Mojave River and the Soda Lake country.

Early on the morning of the 8th the wagon train started [the wagons were being sent back in the direction of San Bernardino], and shortly after the packs were ready, and we were on the trail down the river. On the evening of the 10th we made our camp at the lowest point [probably the Camp Cady site] I had reached on the river during my preliminary examination. From here we followed the broad river-bed for 15 miles, having clay bluffs on each side, when we entered the cañon [Cave Canyon–now known as Afton Canyon], which is about seven miles long, and had running water throughout its whole length. The bed of the stream in the cañon is from 100 to 150 feet in width. On either side the clay bluffs rise often over 100 feet. They are vertical, and, by the action of the water, the surface has in many places assumed the form of Gothic pillars. The clay presents every variety of beautiful and delicate tints–purple, pink, blue, yellow, &c. In the cañon we found cane growing, similar to that mentioned as found in Walker's Pass, and large quantities of it had been cut by the Indians. We saw, however, no fresh Indian signs.

Upon emerging from the cañon, we entered a sandy plain, and at once lost all signs of the river-bed. After travelling 13 miles across this plain, we were fortunate enough to find a hole containing water, and here we made our camp late at night. The water was barely sufficient for our nearly exhausted animals, and a long time was occupied in giving them a scanty supply. This plain had an abundant growth of mezquite trees. We afterwards found some uninhabited Indian huts near here, and saw an abundance of old Indian tracks, but no Indians.

In the morning, upon taking a survey of our position, we found we were near the centre of an irregularly shaped plain, surrounded by hills. To the southeast appeared an opening, and here we concluded was the outlet through which the Mohave continued its course. We went in this direction about thirteen miles [they were heading to the southeast of Soda Lake into the Devil's Playground in the direction of the Kelso Dunes], the first seven or eight of which were over low sand-hills; but afterwards we travelled in the dry bed of a wash, which we found we were *ascending.* Being convinced that we had left the Mohave, and the men who had been sent in search of water having been able to find none, we returned to our camp of the morning. This was another very fatiguing day for the animals, and after reaching camp, there being so little water in the hole, which filled very slowly, it took till 2 a.m. to water them. We observed to-day that to the north of our camp was a large lake-bed [Soda Lake], and here we inferred the waters of the Mohave were collected. The question now was, whether this lake had an outlet, or whether it was a basin, and the terminus of the Mohave. To ascertain this point, Lieutenant [George] Stoneman and myself started to examine the lake, which was about fifteen miles long, and covered with an incrustation of salt, exceedingly bitter. We, however, returned to camp late, without any positive result. Not being willing to move the command upon an uncertainty, I resolved to devote one more day to examination. To the east of our camp was a high range of barren mountains, its crest from fifteen to twenty miles distant [Old Dad Mountain]. Lieutenant Stoneman and myself ascended to the summit of one of the nearest peaks, from which we had an extended view. To the eastward were to be seen nothing but mountains; we saw, however, that to the northward of the salt lake, and not far distant from it, were several other lake-beds. Our view to the southward was bounded by mountains.

The result of these examinations was, that if the Mohave flowed beyond the salt lake, it could flow in no other than a northerly direction through these lake-beds, and the only thing to be done was to proceed in this direction, though directly contrary to the one we wished to go.

We had found at the base of the hills, on the edge of the salt lake, several fine springs, slightly brackish but not unpalatable. Around these was good grass. The camp was moved here, and the animals were refreshed by once more having as much to eat as they wanted [this must have been Soda Springs and if so this may be the first written description of the site].

On the morning of November 16, at 5 o'clock, we started by fine moonlight and travelled to the northern extremity of the salt lake [near present Baker, California], and thence on to the next one [Silver Lake]. We found the two connected by a ditch, cut by water in the clay soil, and about twenty feet wide, with banks two feet high. The two lakes were from three to four miles apart. The second one was six miles long and three broad. The character of the second lake was entirely different from that of the first. It was a dry, hard clay-bed, on which the shoes of the mules scarcely made an impression; while the other was covered with salt, and in many places too soft to travel over. The bases of the hills on the west, and the mountains on the east, were immediate on the lake bank, and as we crossed it we could see there was no outlet in either of these directions; but to the northward the hills were low, and we expected to find here a passage where water could flow. On arriving at the north end of the lake, we found a very low ridge, connecting the hills on either side. We searched for a passage through this ridge, but could find none; but everywhere saw, in the gullies, that the water from rains flowed towards the lake. I hence concluded that the true sink of the Mohave river was in the salt lake, and that the second lake was formed principally from water flowing from the surrounding hills after heavy rains; but that in time of very high water in the salt lake, its surplus flowed through the ditch before mentioned into it.

We crossed this ridge, and at once descended into another valley some two hundred feet lower than the bed of the lake. After travelling four or five miles, we came suddenly upon a wagon-road. We knew it could be no other than the old Spanish trail, and this at once afforded proof that the Mohave river of the maps is a fiction. It was universally supposed by emigrants and others, that when the Spanish trail left the Mohave above the cañon, it never was on it again. The valley in which we struck the trail extended to the northward twenty or thirty miles, bounded on all sides by mountains.

We were now, according to our estimates, over 100 miles in a direct line from the Colorado, with a mountainous country between, and neither wood, water, nor grass that we knew of. To attempt to reach that river would have been madness. Our only alternative was to turn back. We took the wagon-road, and at midnight reached Agua de Tomaso [Bitter Springs], having travelled fifty-five miles without water.

Agua de Tomaso, (or, as I am told, more properly Agua de Tio Mes,) is simply a spring of bitter water, which does not flow more than a hundred yards, and is quite different from what maps generally represent it to be; for they often make it a stream 20 miles long. There was not enough grass here to subsist our animals, and we were, therefore, obliged to push on early in the morning to where the trail strikes the Mohave. From here we went by the usually travelled roads through the settlements.

Instructions of the Secretary of War to Edward F. Beale for the Survey of a Wagon Road from Fort Defiance to the Colorado River.

War Department
Washington April 22d 1857.

Edward Fitzgerald Beale, Esq.
Washington City.

Sir.

Under the provisions of an Act of Congress approved February 17, 1857 you are hereby appointed to construct a wagon road from Fort Defiance to the Colorado river at a point opposite the supposed mouth of the Mohave river.

You will, accordingly, proceed to New Mexico with the least possible delay and organize a working party, to consist of not less than thirty five nor more than fifty five men, whom you will supply with the requisite tools, &c.

You will employ such of your party here as you may think necessary and their traveling expenses will be allowed them to the point of their departure.

You will take a sufficient number of strong wagons, not exceeding ten, to convey the provisions, tools, camp equipage, &c, and to demonstrate practically that the road may be passed over with heavy wagons.

You will be provided with funds at New York, where you will purchase such supplies as cannot be as well obtained upon the frontier or in New Mexico, or elsewhere.

Having organized your party at the most convenient point, you will move to Fort Defiance and commence the construction of the road, which you will make passable for loaded wagons throughout. Especial care will be taken to make all watering places and all fords easily accessible to wagons. Careful estimates of the cost of building bridges at suitable points, will be made to form the basis of appropriations, should the Government find it expedient to construct them.

Having opened the route to the Colorado river, you will, if the funds admit it, pass over the road a second time, improving those portions that most require it.

Should your supply of provisions be short upon your arrival at the supposed mouth of the Mohave river, you will proceed to the settlements at or near Fort Tejon for additional supplies, taking care to travel along the route best adapted to the passage of troops from the Colorado river to Fort Tejon, and to make your trail useful for such purpose.

You will be furnished with copies of the printed maps and reports of the explorations made under this Department of the region through or near which the road will pass.

To aid you in this work you are authorized to employ several assistants, as one topographer or draughtsman, whose compensation will not be above twenty five hundred dollars, one physician and one surveyor at compensations not to exceed five dollars per day, with subsistence, and such other assistants, not to exceed three in number as the service may require, whose compensations shall be at the rate of from five to three dollars per day, with subsistence and transportation. Your own compensation will be at the rate of three thousand dollars per year and necessary traveling expenses. You are likewise authorized to purchase such surveying instruments as the nature of the operation requires.

Upon your arrival at Fort Defiance, an escort of twenty five mounted men, under the command of a Sergeant and two corporals, will be furnished you by the officer in command of that post.

In passing through a country inhabited by hostile Indians, you will carefully guard your party from danger, and omit no occasion to conciliate them by such presents and kind acts as may be within your power.

To enable you to execute these instructions, you will be furnished with the whole sum appropriated by Congress, (fifty thousand dollars,) and your whole expenditures must, in no event, exceed that sum. You will be careful to take suitable vouchers for your disbursements and render them promptly to the Department for transmission to the proper auditing officers.

You will receive an order on Capt. J. N. Palmer to transfer to you as many camels and dromedaries as you may deem necessary, not exceeding twenty five in number, and which you will use on this exploration to test their usefulness, endurance and economy, and as many of the men who accompanied them from the Mediterranean, as you may require.

Lateral explorations will be made whenever you may deem them necessary to develop the resources of the route upon which you will be at work.

In the event of your death, Mr. G. H. Heap, your first Assistant, will continue the duty you have been sent to perform, and will be recognized as your successor in the command of the expedition [Heap did not accompany the party to California].

As the success of the expedition will depend on the condition of your animals, you will furnish them with full forage, whenever it may be necessary.

You will report your progress to the Department fully and frequently, and at the termination of your service, will submit a general report upon the whole subject.

Very respectfully, Yr obt Servt
John B. Floyd, Secretary of War.

Extracts from the Los Angeles Southern Vineyard *of November 26, 1859, pertaining to the first train sent by Captain Winfield Scott Hancock to Fort Mojave under Wagonmaster Joseph Winston in October and November of 1859.*

Los Angeles Route to Fort Mojave.

The great expense of forwarding supplies to Fort Mojave by the circuitous way of the Gulf of California and the Colorado River, as well as the occasional difficulties of navigating that stream, induced Capt. W. S. Hancock, of the Q. M. Department, U. S. Army, to make another effort to send supplies by land from San Pedro, the port of this city, to Fort Mojave.

Judge Winston left here Oct. 21, with 25,165 pounds of freight, and 250 feet pine lumber, all of which he delivered in good order at the Fort on the 5th of Nov. Besides the freight there was 10,100 pounds barley for forage for the teams.

The cost of transportation from San Francisco to the Fort by this route has been about 4½ cents per pound, including freight and all incidental expenses, except the use of wagons and mules. The shoeing of the mules was one of the items of expense charged. As the cost heretofore by the way of the Gulf and river has been about 15 cents per pound, it will be seen that here is a saving of more than ten cents per pound.

Much credit is due Capt. Hancock for his efforts in demonstrating the advantages of this route. In opposition to the most discouraging reports he felt confident that a cheaper and more expeditious mode of transportation than the river was practicable.

Although the report of Col. Hoffman and officers who accompanied him on his march from this city to the Mojave Villages, and more recently the expedition sent out with government freight by Major Banning, were such as to cast doubt upon the result of Captain Hancock's undertaking he was not discouraged from the attempt. The result has been a complete success and proved the soundness of his judgment. Freight can be taken from San Francisco to Fort Mojave by this road in twenty days and at less than one third the sum paid by the river route.

Itinerary, by Joseph Winston, Esq.

Oct. 21 [1859].–Left Los Angeles 7, A.M., with ten wagons drawn by eight mules each. The load of each wagon on starting was 3600 lbs. Camped at Mud Springs at 5, P.M., distance 28½ miles.

Oct. 22d, at 12, M., camped at Cucamungo, 12½ miles.

On the 23d, to Martin's Ranch, 13 miles, camped at 1, P.M.

24th camped at old house in the canyon, 7½ miles, at 11, A.M. Road sandy and stony.

25th. To foot of mountain, 8¼ miles, road heavy from deep sand, and ascending, camped at 12, M., without water–sent mules back to last camp for water, from whence they returned at sunset.

26th, started at daybreak, doubled teams, taking only four wagons at a time, for want of chains, and moved to the summit, about half a mile, without any difficulty or detention. Broke a few fifth chains only. At 9 o'clock had the train up and moved on to the upper crossing of the Mojave river–road good, camped at 4, P.M., 23¼ miles from the last camp, plenty of good grass.

27th, crossed the river, a beautiful swift running stream of 2000 cubic inches volume, and traveled down the right bank 19½ miles, N.E., with good road to Government reserve–camped at 3, P.M., abundant of good grass, and water standing in large pools. The river sinks about three miles below last camp, occasionally rising and sinking at short distances, making ponds every ten or fifteen miles. Fine cottonwood timber along the banks, and about a section of good land at this camp.

28th, continued down same bank 21 miles nearly east to Fish ponds [near the present Barstow]. At Sugar Loaf Mountain traveled a short distance in the river bed. Road level but sandy in places.

29th; soon after leaving camp crossed the bed of the river, and at 12 miles, left the Salt Lake road, which here leaves the Mojave. Continued down the river, east, to Lieut. Davis' depot [the site of future Camp Cady], 23 miles, camped at 4, P.M. The road generally good, but in some places sandy. Plenty of wood, water and grass. Considerable extent of good pasture land. One or two hundred tons of good hay could be cut here annually.

30th, continued down the river, about east, to the Cave, 15½ miles. Road level, but sandy with rocks. Water and grass plenty. At this place the river rises and runs some six or seven miles.

Oct. 31st, kept on down the river through a gorge [the present Afton Canyon], the mountains rising on both sides from two to four hundred feet in perpendicular clifts. The pass is from two to three hundred yards wide and four miles long. Coming out of the gorge the road enters upon a large plain of coarse sand, with some dwarf grease wood. Here the Mojave river sinks. Its waters however, find their way over the plain in time of floods to Soda Lake where we camped at 4, P.M., 19 miles. Course about N.E., road sandy. At this place there are some five or six acres of rushes and salt grass about the springs, which furnish a large supply of water. There is also plenty of galeta grown near this place.

Nov. 1, Remained in camp until 4, P.M., then started and traveled 17 miles N.E., and camped at 1, A.M., of the 2d, opposite the first old crater without water or grass. Road heavy sand [the old crater referred to is one of the cinder cones southeast of present Baker, California].

2d. Left camp about sunrise, and traveled nearly east, with sandy road, 12½ miles, to Marl Springs, where we found plenty water and some grass, about one mile S.E. of the spring.

3d. Good road, except about three miles of sand for 19½ miles, N.E., to Rock Spring, where water, grass and cedar wood are all plenty.

4th. Traveled 24 miles East with good road except last mile [which was over Pah-Ute Hill] to Pah Utah Creek. Water and grass in abundance.

5th. Reached the Colorado river opposite the Fort, at 3, P.M. Distance, 22 3/4 miles good road with plenty grass on the river. Whole distance from Los Angeles, 286 miles.

Removed to the Fort on the 5th, and left on the morning of the 7th, and reached San Bernardino on the 17th at 11, A.M., without the loss of a mule or the breaking of a wagon. No goods were unloaded during the trip, and teams were not doubled but once, and then in the Canyon of San Bernardino [Cajon Pass]. Mules all returned in good working order.

The train employed ten teamsters and two herders, at $30 per month each. It was absent from this city thirty-three days. Two days were added by going to San Bernardino for loads for Los Angeles.

Report of Capt. Henry M. Enos of his exploration of the road between Forts Whipple and Mojave in December of 1863 and January of 1864.

Albuquerque, New Mexico,
February 15th 1864.

Captain B. C. Cutler,
Assistant Adjutant General.
Head Quarter's Dept. New Mexico
Santa Fe N. M.

Sir:

By Par. 4 of General Orders No. 27 dated Head Qrs. Dept. New Mexico, Santa Fe, N.M., Oct. 24, 1863, I was detailed as one of the

Board of Officers to select and fix the exact site and submit a plan, with an estimate of costs, for the Post of Fort Whipple, which was by the same order, to be established at or near the new gold fields in the Territory of Arizona; and by Par. 7 of this order, I was ordered, after Fort Whipple had been established, to examine the country westwardly from that post with a view to finding a practicable wagon road, to some point on the Colorado river, if possible, below Fort Mojave, and to report in detail upon such road as to its distance character, &c.

In compliance with the order above referred to, I left the Rio Grande on the 11th of November, 1863, with an escort of five men of Company "D" 1st. Cavalry Cal. Vols., and on the 19th came up with Major Willis' command, which was then en route to the new gold fields, at Jacob's Well. I traveled with this command and its unwieldy supply train, drawn by oxen until the 30th, when Major Willis determined to go in advance with the greater part of the cavalry detachment, leaving the Infantry and supply train to follow. We reached the present site of Fort Whipple on the 10th of December.

While waiting at Fort Whipple for the remainder of Major Willis' command to arrive, I made enquiries relative to the routes which were then being traveled to the mines from the Colorado river. There were three; one from La Paz via William's Fork; one from Fort Mojave, which follows Beal's wagon road for about forty miles, and another from Fort Mojave, striking and crossing the William's Fork below the mouth of the Santa Maria river. This latter road was represented as being wholly impracticable for wagons, and not a good packtrail, whatever its course it must be considerable longer than either of the others. I could hear of no one having traveled it a second time.

I was unable to procure a guide who was sufficiently well acquainted with the country–Capt. Joe Walker, the only man in the mines whose general knowledge of it could be relied upon, was absent. I however procured the services of a man who had been over the William's Fork route, and who had some experience as a mountaineer.

Major Willis furnished me with an escort of one non Com. Officer and nine men of Company "D" 1st Cav. Cal. Vols. and one wagon and team of 8 mules, and on the 23d of December, I left Fort Whipple with a view of striking the Colorado at the mouth of William's Fork.

About 14 miles from the post I crossed the trail leading from Fort Mojave to the mines, and at about 20 miles crossed by easy ascent and descent the divide which separates the waters of the San Francisco and Santa Maria rivers, the road passing between the Aztec range and the Black or Granite mountain. After crossing the divide I followed a tributary of the Santa Maria, reaching this stream about 42 miles from the Fort, the general direction traveled being southwest, as near as I could judge. I came upon the trail from La Paz about 3 miles before reaching the Santa Maria. I followed the river for 1½ miles, which at this point runs nearly west, then leaving it to my right, kept upon the trail to where it enters a deep and narrow cañon. This cañon is impassable for a loaded wagon. With considerable labor it could be cleared of the rocks which now prevent its passage with wagons, and there is but one heavy grade in it to reach the high table land or mesa. The top of the mesa can also be reached by ascending a steep slope to the right and west of the cañon, but as the course would have taken me over about four miles of volcanic rock, some of which is almost impassable and which would risk the breaking of my wagon, and as I had no means of repairing damages should any occur, I determined to return and try the road to Fort Mojave. I had up to this point found a fair wagon road, with abundance of water at convenient distances; grass, as a general thing, was plenty, fuel easily obtained. I was, according to my guide's calculation within 125 miles of the Colorado river. He represented the country to be with a few exceptions good for a wagon road until reaching the William's Fork. After reaching the cañon of this stream the road would be heavy from sand, grass poor in quality, drift wood furnishing fuel, the latter is represented as lying as high as forty feet above the bed of the stream in some of the cañons through which it flows. The country on either side is rough and mountaneous. The Santa Maria river, a few miles below the point where the trail leaves it enters a deep and narrow cañon, almost impassable, for man or beast.

On the 29th of December I returned to the point where I had crossed the road from Fort Mojave. A few light wagons had previously passed over this route which is known as the Irrataba trail, a celebrated Mojave chief of that name having been the first to conduct emigrants over it with wagons. It was represented by those who had traveled it as being an excellent one for wagons, with an abundance of wood, water, and grass. From this point, until reaching what Lieut. Whipple calls Pueblo creek, a distance of 20 miles, the road passes over a mountaineous country, covered with a thick growth of cedar. The road follows Pueblo creek to its source, passing through Aztec pass, then down William's Fork to near the base of Cross mountain. After passing this mountain the road bears to the south towards Aguarias mountain, crossing it to the north of the point where the White Cliff Creek finds a passage. Crossing the Big Sandy it follows the north western base of the Cerbat range for about 15 miles then bears to the north, striking Beal's wagon road at a distance of forty miles from Fort Mojave, at a place known to emigrants as Rail Road Spring. From this point I followed Beal's wagon road to Fort Mojave.

The road from Pueblo Creek to Railroad Spring is good with the exception of about one mile at the point where it leaves William's Fork, and about four miles in crossing the Aguarias range. At both of these places a good road can easily be made. Grass is abundant nearly the whole distance. Fuel is plenty, till after leaving the Aguarias mountain. Water along William's Fork is found in holes, and I believe is permanent and in sufficient quantities for the animals of ordinary trains. The longest distance from water to water, 25 miles, is between White Cliff Creek and the Hualpais Spring, which is situated on the north western slope of the Cerbat range.

From Railroad Spring to Fort Mojave the country is almost a desert. A few Mesquite shrubs, Cactu's Spanish Bayonet, and sage brush, constitute nearly all the vegitation seen, except in the gorges of the mountains where a few small cedars may be found. The road over the Blue ridge has very heavy grades and in places is quite rough [the reference here is to Sitgreaves Pass through the Black Mountains]. On both sides it follows deep and Sandy arroyos which bear evidences of carrying much water at times; considerable labor will be required to make this part of the road practicable for heavily loaded wagons.

At Silver creek on the western slope of the Blue ridge, and twelve miles from Fort Mojave, rich gold and silver leads have been discovered [the reference is to the San Francisco Mining District established in 1863]. Companies in San Francisco have been organized to work them. After leaving the Arroyo of Silver creek the road to Fort Mojave is good, except about two miles of sand. By the route I traveled the distance from Fort Whipple to Fort Mojave is 156 miles–see list of distances enclosed–This distance can be shortened from ten to fifteen miles by an expenditure of a reasonable amount of labor. The rough and mountainous country which I passed over before reaching Pueblo Creek can be avoided, and a good road obtained by going direct from Fort Whipple to the mouth of its cañon. I estimated the distance to be about eighteen miles from the Fort to the point where this stream disappears beneath the surface for the last time. This, like all other streams in that portion of Arizona, rises and sinks at intervals. The labor of fifty men for sixty days would make the road from Fort Mojave to Fort Whipple practicable for six mule teams loaded with from 3000 to 3500 pounds. The road, as I found it compares favorably with the mountain roads in the northern part of New Mexico.

I arrived at Fort Mojave on the 10th of January 1864. I had hoped at that post to find forage for my animals, which had become poor and weak, on account of the long distance they had traveled without rest, and a severe snow storm that fell on the 5th of January; but I could procure nothing, except the poorest kind of bottom hay.

I had intended to return to Fort Whipple with packs via the mouth of William's Fork but owing to the poor condition of my animals, the want of means of improving their condition, or exchanging them for better ones, I determined to leave my wagon and return with packs by the route I had traveled. On my return I followed Pueblo creek to the mouth of its cañon, which runs nearly east, opening into the "Val de Chino," about 14 miles to the north west of Fort Whipple.

The Colorado river is said to be navigable with boats of light draft, carrying from fifty to one hundred tons, for seven or eight months of the year. There is but one boat on the river that runs above Fort Yuma, and this only as occasion demands. The cost of transportation from San Francisco to Fort Mojave by sea and river is seven and one-half cents (7½) per pound. The cost from Los Angeles to the same point is twelve cents (12) per pound. At the latter rate it would cost to carry freight from Fort Mojave to Fort Whipple about five cents (5) per pound.

The navigation of the river is as yet controlled by one company. It is believed that the amount of freight now awaiting transportation at the mouth of the river, and at Fort Yuma, and the prospects of an increasing demand for transportation, will warrant the building of other and more suitable boats for navigating this stream. A competition will tend to reduce the present high rates of freight. I was informed that boats were being built to be put on the river at as early a day as possible.

That portion of Arizona over which I traveled can never amount to much as an agricultural country. A small strip of land not over one thousand acres, bordering the Santa Maria can be cultivated by means of irrigation, and also about an equal amount along Pueblo creek. These are the only two places that I saw, where, I think, farming will ever pay. Although the country abounds with the finest quality of grass, the want of water will prevent its becoming a very extensive grazing country. There is however sufficient to sustain large herds of horses, sheep and cattle.

The Indians which inhabit this portion of Arizona are known by the various names of Tonto Apaches, Tampais, and Hualpais. They have a few horses, and live upon cedar berries, piñon nuts, and game. I saw some places where they had raised small patches of corn the previous summer.

Time alone will show whether the mineral wealth of Arizona will prove to be as great as is anticipated. The development of this country is retarded by the hostile attitude of the Indians which infest the region between the present mining districts and the Gila river, and by the difficulty in getting supplies.

The removal of these impediments depends greatly upon what action the Government may take in the matter. Should those in authority feel justified from the developments already made, to send troops to rid the country of Indians, and take means to improve the roads over which provisions and supplies of all kinds must be carried, I believe it will not be many months before the world will be astonished at the riches which lie buried in the mountains of this unexplored and undeveloped portion of our public domain.

Respectfully submitted
H. M. Enos
Capt. & A. Q. M.

NOTES

In general, these notes have been assembled to establish the sources of quoted material. The broad scope of this book precludes extensive references. Earlier books in the *Tales of the Mojave Road* series contain more detailed references to source material and, in several instances, detailed descriptions and discussions of places and people mentioned in this present book. Subsequent monographs in the series will treat other aspects of the history of the Mojave Road and provide additional references.

1. In this book the spelling "Mohave" is used for the Indian Tribe, most names along the Colorado River, and the Indian trail that crossed the eastern Mojave Desert. The fort that was established at Beale's Crossing in 1859, the desert and river in California, and the wagon road are spelled with the "j," i.e. Mojave. This convention seems to be as consistent as possible with historical and current practice.

2. R. B. Stratton, *Life Among the Indians or: The Captivity of the Oatman Girls among the Apache & Mohave Indians* (San Francisco: Grabhorn Press, 1935), 106.

3. For an excellent article on the Mojave River see the following: Edmund C. Jaeger, "The Upside Down Mojave River," *Desert Magazine,* June, 1957, 18-20.

4. John Galvin, ed., *A Record of Travels in Arizona and California 1775-1776* (San Francisco: John Howell Books, 1965), 1.

5. These quotations are found in *Ibid,* 35, 37, 44, 57, 59.

6. *Ibid,* 40. An earlier translation of the journal kept by Garces is available in Elliott Coues, *On the Trail of a Spanish Pioneer: Garces Diary 1775-1776* (2 Vols.; New York: Francis P. Harper, 1900).

7. George W. Beattie, trans., "Diary of Fr. Joaquin Pasqual Nuez, Minister of San Gabriel and Chaplain of the Expedition Against the Mohave Indians, Begun by Lieutenant Gabriel Moraga, November, 1819," *San Bernardino Museum Association Quarterly,* II (Winter, 1955), Appendix III.

8. Smith to Clark, Little Lake of Bear River, 12 July 1827 (Clark Papers, Kansas State Historical Society).

9. Maurice S. Sullivan, *The Travels of Jedediah Smith* (Santa Ana: The Fine Arts Press, 1934), 31-33.

10. Timothy Flint, ed., *The Personal Narrative of James O. Pattie* (Philadelphia and New York: J. B. Lippincott Co., 1962), 85-87.

11. Blanche C. Grant, ed., *Kit Carson's Own Story of His Life* (Taos: 1926), 14-15.

12. *Ibid,* 18-19.

13. Alice Bay Maloney, "Peter Skene Ogden's Trapping Expedition to the Gulf of California, 1829-1830," *California Historical Society Quarterly,* XIX, No. 4 (December, 1940), 312.

14. LeRoy R. Hafen and Ann W. Hafen, *Old Spanish Trail–Santa Fe to Los Angeles,* The Far West and Rockies Historical Series, Vol. I (Glendale: The Arthur H. Clark Co., 1954), 81.

15. *Ibid*, 156-157.

16. Wilmington *Journal*, 20 Oct. 1866.

17. The Fremont quotations can be found in the following: Capt. John Charles Fremont, *Report of the Exploring Expedition to the Rocky Mountains in the Year 1842, and to Oregon and North California in the Years 1843-44* (Washington, D.C.: Printed by Order of the Senate, 1845), 258-261.

18. James H. Simpson, *Journal of a Military Reconnaissance, from Santa Fe, New Mexico, to the Navajo Country, made with the Troops under Command of Brevet Lieutenant Colonel John M. Washington, Chief of Ninth Military Department, and Governor of New Mexico, in 1849* (Philadelphia: Lippincott, Grambo and Co., 1852), 124-125.

19. U. S. Congress, Senate, "Report of an Expedition Down the Zuni and Colorado Rivers," *Senate Exec. Doc. 59,* 32 Cong., 2 Sess., 4.

20. The Aubry journals are available in the following: Ralph B. Bieber, ed., *Exploring Southwestern Trails 1846-1854,* The Southwest Historical Series, Vol. VII (Glendale: The Arthur H. Clark Co., 1938), 353-383.

21. *Ibid,* 353.

22. Donald Chaput, *Francois X. Aubry–Trader, Trailmaker and Voyageur in the Southwest 1846-1854* (Glendale: The Arthur H. Clark Co., 1975), 149.

23. *Ibid,* 157.

24. The Williamson survey report is the following: U. S. Congress, House, "Reports of Explorations and Surveys, to Ascertain the Most Practicable and Economical Route for a Railroad from the Mississippi River to the Pacific Ocean," *House Exec. Doc. 91,* 33 Cong., 2 Sess., Vol. V (Washington: 1856), 32-34.

25. Baldwin Mollhausen, *Diary of a Journey from the Mississippi to the Coasts of the Pacific* (2 Vols.; London: 1858), Vol. 2, 241-244. Mollhausen reports in detail on the abandonment of Whipple's wagons.

26. U. S. Congress, House, "Reports of Explorations and Surveys, to Ascertain the Most Practicable and Economical Route for a Railroad from the Mississippi River to the Pacific Ocean," *House Exec. Doc. 91,* 33 Cong., 2 Sess., Vol. III (Washington: 1856), 119. This is the Whipple Survey report.

27. *Ibid,* 123.

28. David S. Stanley, *Diary of Lt. David Sloan Stanley, 2nd U. S. Dragoons of an Expedition Which Made a Journey Overland from Fort Smith, Arkansas–to San Diego, California* (Unpublished Manuscript, Library of Congress, 1935?), 33.

29. Whipple, "Reports of Explorations and Surveys...," 20.

30. For an excellent treatment of this general subject see the following: William H. Goetzmann, *Army Exploration in the American West* (New Haven: Yale University Press, 1959).

31. I have not quoted any major part of the material from the Whipple expedition in this book. The reason for this is that I have hopes that all of this extensive material can be brought together in a future issue in the *Tales of the Mojave Road* series.

32. Secretary of War, Letters Sent, Military Affairs (National Archives, M-6, Roll 39).

33. See, for example, Harlan D. Fowler, *Camels to California* (Stanford: Stanford University Press, 1950) or *Uncle Sam's Camels* as cited in footnote 34.

34. Lewis Burt Lesley, ed., *Uncle Sam's Camels, the Journal of May Humphreys Stacey Supplemented by the Report of Edward Fitzgerald Beale (1857-1858)* (Cambridge: Harvard University Press, 1929), 143.

35. In discussing difficulties with the Mohaves in 1857 and 1858 I have referred to "Mormons" and "Americans" as though they were separate groups. In fact, "Mormons" were "Americans" and my usage of the terms could be considered technically incorrect. In the vernacular language of the day, however, both groups—that is, "Mormons" and "other Americans"—used the terms as I am using them here.

36. Burton to Cooper, San Francisco, 11 August 1860 (National Archives, M-567, Roll 620). Col. S. Cooper was Adjutant General of the Army at the time.

37. George Alonzo Johnson, "The Steamer General Jesup," *Quarterly of the Society of California Pioneers,* IX (June, 1932), 110-111.

38. It should be noted that not all of Blake's command accompanied him to the Colorado River. To streamline his party thereby permitting him to keep pace with Beale, Blake established a temporary camp at the head of Cave Canyon. He left his wagons and part of his troops there as a guard. For the next several years this point was known as "Blake's Camp." After about 1861 the name dropped out of use.

39. Lesley, *Uncle Sam's Camels,* 261-262.

40. Arthur Woodward, *Feud on the Colorado* (Los Angeles: Westernlore Press, 1955), 100-101.

41. Los Angeles *Star,* 20 Feb. 1858.

42. U. S. Congress, Senate, "Report upon the Colorado River of the West Explored in 1857 and 1858 by Lieutenant Joseph C. Ives," *Senate Exec. Doc.,* 36 Cong., 1 Sess., 89.

43. Ives to Mackall, 22 Mar. 1858 (National Archives, RG-393, Department of California Letters Received).

44. A typescript of the Lyman Journals is available in the Special Collections in the Brigham Young University Library; Provo, Utah.

45. The complete subject of Mormon presence along the Colorado River and among the Mohave Indians during the period of interest here is treated in the following: Melvin T. Smith, "Colorado River Exploration and the Mormon War," *Utah Historical Quarterly,* XXXVIII (Summer, 1970).

46. Smith to Hon. A. Campbell, 8 Feb. 1858 (National Archives, M-95, Roll 16).

47. Robert Glass Cleland, *The Cattle on a Thousand Hills* (San Marino: The Huntington Library, 1964), 264-265.

48. *Ibid,* 268-269.

49. A.G., U. S. Army, to Clarke, 1 Dec. 1858 (National Archives, RG-393, Department of California Letters Received).

50. Hoffman to Mackall, 16 Jan. 1859 (National Archives, RG-393, Department of California Letters Received).

51. Hoffman to A.A.G. Department of California, 19 Jan. 1859 (National Archives, RG-393, Department of California Letters Received).

52. Hoffman to A.A.G. Department of California, 27 Jan. 1859 (National Archives, RG-393, Department of California Letters Received).

53. Mackall to Tipton, 1 Feb. 1859 (National Archives, RG-393, Department of California Letters Sent).

54. Clarke to A.G. U.S. Army, 21 Mar. 1859 (National Archives, RG-393, Department of California Letters Sent).

55. Hoffman to Mackall, 24 Apr. 1859 (National Archives, RG-393, Colorado Expedition Letters Sent).

56. Hoffman to Mackall, 18 May 1859 (National Archives, RG-393, Colorado Expedition Letters Sent).

57. The subject of the theft of Bishop's caches at Pah-Ute Creek is treated in detail in the following: Dennis G. Casebier, *Fort Pah-Ute California* (Norco: Tales of the Mojave Road, 1974).

58. A.G. U.S. Army to Clarke, 30 Apr. 1859 (National Archives, RG-393, Department of California Letters Received).

59. Contemporary accounts report the majority of Mohave hostages as having been killed in the escape. However, in a reminiscence by a Mohave given to A. L. Kroeber in 1903 it is stated that only two Mohaves were killed. See: A. L. Kroeber and C. B. Kroeber, *A Mohave War Reminiscence, 1854-1880* (Berkeley-Los Angeles-London: University of California Press, 1973), 25-27.

60. Armistead to Mackall, 31 Aug. 1859 (National Archives, RG-393, Department of California Letters Received).

61. Los Angeles *Southern Vineyard,* 26 Nov. 1859.

62. Los Angeles *Star,* 26 Nov. 1859.

63. The story of this wagon train was told by George William Beattie and Helen Pruitt Beattie in their excellent book *Heritage of the Valley.* They correctly assess the crucial importance of the train in terms of the lack of supplies at Fort Mojave and point out the inability of steamers to supply the fort regularly. Also, they mention the difficulties created by Hoffman's evaluation of the Mojave Road. They then go on to tell a tale wherein a technical error is committed and, while not of great importance, this error has been repeated so many times in subsequent literature that I feel justified in taking the space here to point it out. The error results from the assumption on the part of the Beatties that Major Armistead was still at Fort Mojave when Hancock's train arrived. The much quoted and copied paragraph reads as follows: "*The Major Armistead to whom Captain Hancock sent relief was one of the Southerners who resigned his commission in the Union army at the outbreak of the Civil War in order to join the Confederate forces, and he was the Confederate General Armistead who was mortally wounded in the*

terrible charge at Gettysburg. Captain Hancock was the Union Major-General whose men faced Armistead there. It was one of the ironies of fate that Hancock should have saved him once, only to destroy him four years later on the opposite side of the continent."

The tragedy the Beatties are pointing out is very real. Armistead and Hancock had been brother-officers in the 6th Infantry since 1844. Furthermore, they were close friends. And, it is a terrible truth that they faced each other at Gettysburg and that Armistead was killed in Pickett's famous charge against Hancock's position. In fact, the personal tragedy for these people was even broader than the account in *Heritage of the Valley* tells. There was a third 6th Infantry officer and former Fort Mojave commander and friend of Hancock and Armistead at Gettysburg on July 3, 1863–Richard Brooke Garnett. He, too, was a brigadier general in the Confederate Army and he too laid down his life for the Lost Cause that day. So there is an abundance of personal tragedy to write about here. The problem with the account in the Beatties' book is that Armistead was not at Fort Mojave when Joseph Winston arrived and he had not been there for more than two months. At the time Hancock was supposed to be saving him on the west side of the continent, Armistead was *simply not there.* He was in his home state of Virginia enjoying a well-earned, and lengthy, leave of absence. He had left Fort Mojave on September 3, 1859, on the river steamer *Cocopah* and traveled on it down to Yuma. From there he took the overland mail stage to Los Angeles (where he visited with Hancock) and then he went on to San Francisco where he boarded an ocean steamer for the east. Fort Mojave was probably the furthest thing from his mind on November 5, 1859, when Hancock's train arrived at Fort Mojave.

64. The resulting campaign against the desert Indians is treated in detail in the first book in this series: Dennis G. Casebier, *Carleton's Pah-Ute Campaign* (Norco: Tales of the Mojave Road, 1972).

65. Almira R. Hancock, *Reminiscences of Winfield Scott Hancock* (New York: Charles L. Webster & Co., 1887), 60-62.

66. Pauline Henson, *Founding A Wilderness Capital: Prescott, A.T. 1864* (Flagstaff: Northland Press, 1965), 170.

67. Santa Fe *Gazette,* 2 Jul. 1864.

68. The story of this fight is treated in detail in the following: Dennis G. Casebier, *The Battle at Camp Cady* (Norco: Tales of the Mojave Road, 1972).

69. Prescott *Arizona Miner,* 30 Nov. 1866.

70. The stories of the outposts along the California stretch of the road are available in the following: Dennis G. Casebier, *Camp Rock Spring California* (Norco: Tales of the Mojave Road, 1973).

71. Prescott *Arizona Miner,* 9 Nov. 1867.

72. San Bernardino *Guardian,* 28 Aug. 1875.

INDEX

Other Books

by

Dennis G. Casebier

Camp El Dorado, Arizona Territory

Carleton's Pah-Ute Campaign

The Battle at Camp Cady

Camp Rock Spring, California

Fort Pah-Ute, California